Endurance

Endurance

A Year in Space, A Lifetime of Discovery

SCOTT KELLY

with Margaret Lazarus Dean

Doubleday

LONDON • TORONTO • SYDNEY • AUCKLAND • JOHANNESBURG

TRANSWORLD PUBLISHERS
61–63 Uxbridge Road, London W5 5SA
www.penguin.co.uk

Transworld is part of the Penguin Random House group of companies
whose addresses can be found at global.penguinrandomhouse.com

First published in Great Britain in 2017 by Doubleday
an imprint of Transworld Publishers

A CIP catalogue record for this book
is available from the British Library.

ISBNs
9780857524751 (hb)
9780857524768 (tpb)

Typeset in 11/15.25pt Minion
Printed and bound by Clays Ltd, Bungay, Suffolk

Penguin Random House is committed to a sustainable
future for our business, our readers and our planet. This book
is made from Forest Stewardship Council® certified paper.

1 3 5 7 9 10 8 6 4 2

To Amiko,

with whom I've shared this journey

A man must shape himself to a new mark directly the old one goes to ground.

—SIR ERNEST SHACKLETON,
Antarctic explorer and captain of the Endurance, *1915*

Endurance

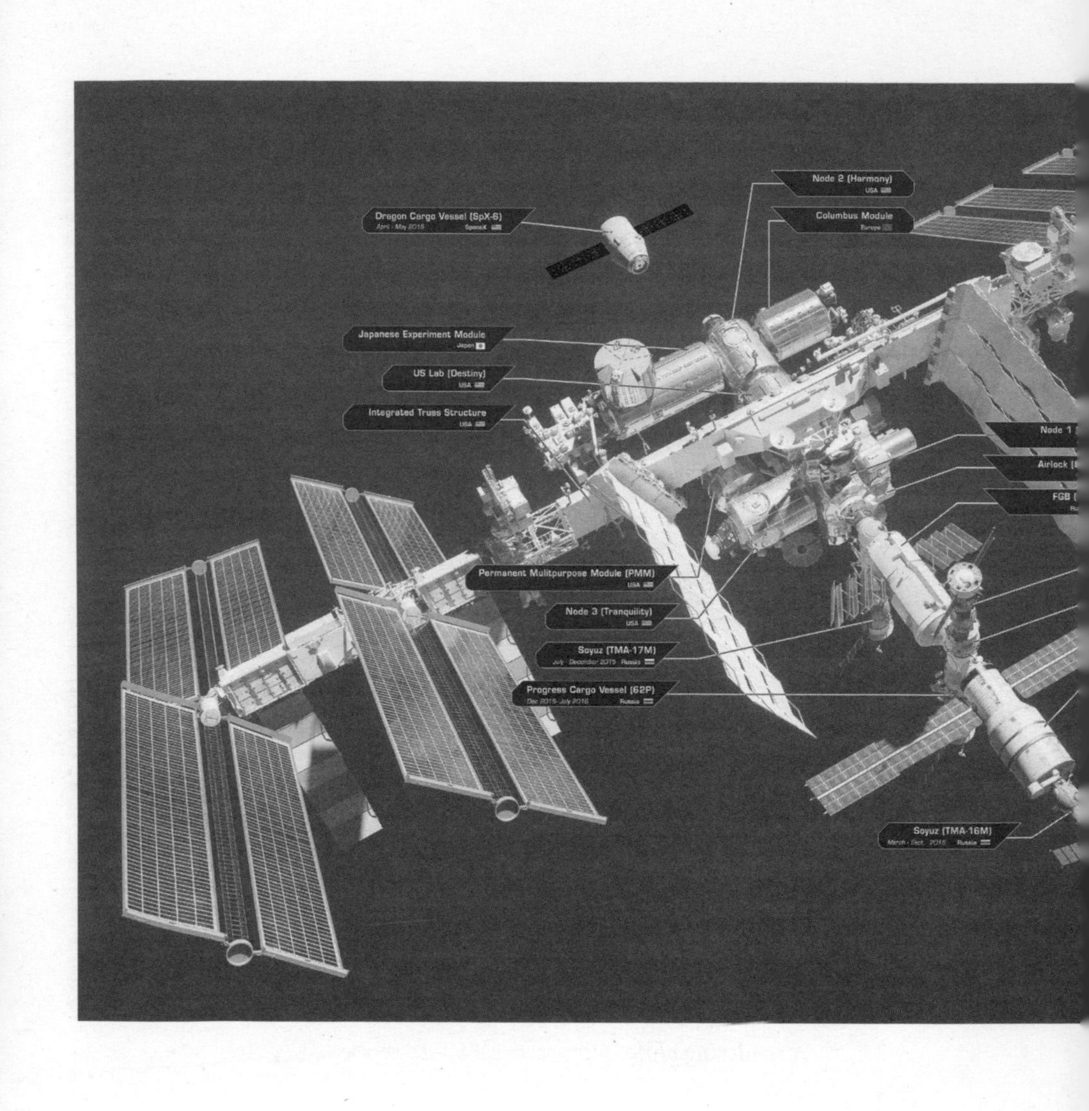

Dragon Cargo Vessel (SpX-6)
April - May 2015
SpaceX
Node 2 (Harmony)
USA
Columbus Module
Europe
Japanese Experiment Module
Japan
US Lab (Destiny)
USA
Integrated Truss Structure
USA
Node 1
Airlock
FGB
Permanent Mulitpurpose Module (PMM)
USA
Node 3 (Tranquility)
USA
Soyuz (TMA-17M)
July - December 2015
Russia
Progress Cargo Vessel (62P)
Dec 2015- July 2016
Russia
Soyuz (TMA-16M)
March - Sept. 2015
Russia

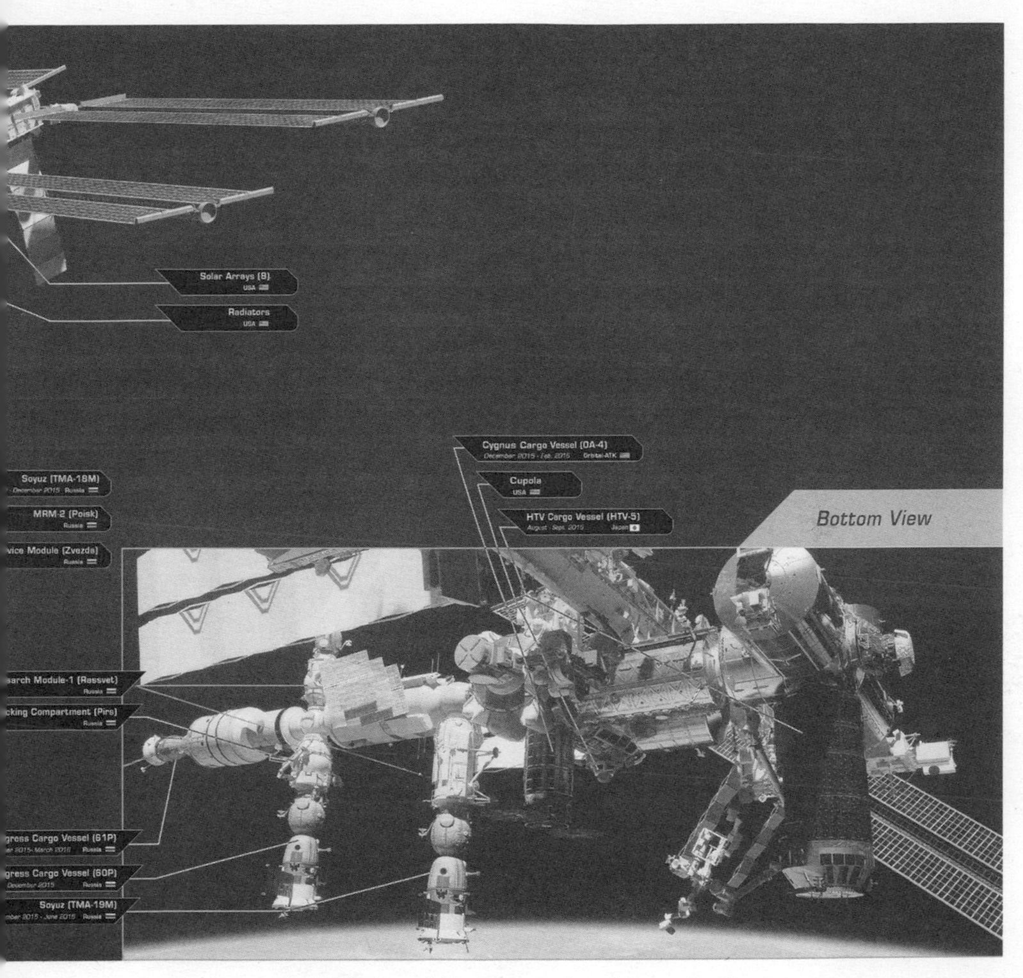

A rendering of the International Space Station

Prologue

I'M SITTING AT the head of my dining room table at home in Houston, finishing dinner with my family: my longtime girlfriend, Amiko; my daughters, Samantha and Charlotte; my twin brother, Mark; his wife, Gabby; his daughter, Claudia; our father, Richie; and Amiko's son, Corbin. It's a simple thing, sitting at a table and eating a meal with those you love, and many people do it every day without giving it much thought. For me, it's something I've been dreaming of for almost a year. I contemplated what it would be like to eat this meal so many times, now that I'm finally here, it doesn't seem entirely real. The faces of the people I love that I haven't seen for so long, the chatter of many people talking together, the clink of silverware, the swish of wine in a glass—these are all unfamiliar. Even the sensation of gravity holding me in my chair feels strange, and every time I put a glass or fork down on the table there's a part of my mind that is looking for a dot of Velcro or a strip of duct tape to hold it in place. I've been back on Earth for forty-eight hours.

I push back from the table and struggle to stand up, feeling like an old man getting out of a recliner.

"Stick a fork in me, I'm done," I announce. Everyone laughs and encourages me to go and get some rest. I start the journey to my bedroom: about twenty steps from the chair to the bed. On the third step, the floor seems to lurch under me, and I stumble into a planter. Of course it wasn't the floor—it was my vestibular system trying to readjust to Earth's gravity. I'm getting used to walking again.

"That's the first time I've seen you stumble," Mark says. "You're doing pretty good." He knows from personal experience what it's like to come back to gravity after having been in space. As I walk by Samantha, I put my hand on her shoulder and she smiles up at me.

I make it to my bedroom without incident and close the door behind me. Every part of my body hurts. All of my joints and all of my muscles are protesting the crushing pressure of gravity. I'm also nauseated, though I haven't thrown up. I strip off my clothes and get into bed, relishing the feeling of sheets, the light pressure of the blanket over me, the fluff of the pillow under my head. All of these are things I missed dearly. I can hear the happy murmur of my family behind the door, voices I haven't heard without the distortion of phones bouncing signals off satellites for a year. I drift off to sleep to the comforting sound of their talking and laughing.

A crack of light wakes me: Is it morning? No, it's just Amiko coming to bed. I've only been asleep for a couple of hours. But I feel delirious. It's a struggle to come to consciousness enough to move, to tell her how awful I feel. I'm seriously nauseated now, feverish, and my pain has gotten worse. This isn't like how I felt after my last mission. This is much, much worse.

"Amiko," I finally manage to say.

She is alarmed by the sound of my voice.

"What is it?" Her hand is on my arm, then on my forehead. Her skin feels chilled, but it's just that I'm so hot.

"I don't feel good," I say.

I've been to space four times now, and she has gone through the whole process with me as my main support once before, when I spent 159 days on the space station in 2010–11. I had a reaction to coming back from space that time, but it was nothing like this.

I struggle to get up. Find the edge of the bed. Feet down. Sit up. Stand up. At every stage I feel like I'm fighting through quicksand. When I'm finally vertical, the pain in my legs is awful, and on top of that pain I feel something even more alarming: all the blood in my body is rushing to my legs, like the sensation of the blood rushing to your head when you do a headstand, but in reverse. I can feel the tissue in my legs swelling. I shuffle my way to the bathroom, moving my weight from one foot to the other with deliberate effort. Left. Right. Left. Right.

I make it to the bathroom, flip on the light, and look down at my legs. They are swollen and alien stumps, not legs at all.

"Oh, shit," I say. "Amiko, come look at this."

She kneels down and squeezes one ankle, and it squishes like a water balloon. She looks up at me with worried eyes. "I can't even feel your anklebones," she says.

"My skin is burning, too," I tell her. Amiko frantically examines me. I have a strange rash all over my back, the backs of my legs, the back of my head and neck—everywhere I was in contact with the bed. I can feel her cool hands moving over my inflamed skin. "It looks like an allergic rash," she says. "Like hives."

I use the bathroom and shuffle back to bed, wondering what I should do. Normally if I woke up feeling like this, I would go to the emergency room, but no one at the hospital will have seen symptoms of having been in space for a year. I crawl back into bed, trying to find a way to lie down without touching my rash. I can hear Amiko rummaging in the medicine cabinet. She comes back with two ibuprofen and a glass of water. As she settles down, I can tell from her every movement, every breath, that she is worried about me. We both knew the risks of the mission I signed on for. After six years together, I can understand her perfectly even in the wordless dark.

As I try to will myself to sleep, I wonder whether my friend Mikhail Kornienko is also suffering from swollen legs and painful rashes—Misha is home in Moscow after spending nearly a year in space with me. I suspect so. This is why we volunteered for this mission, after all: to discover how the human body is affected by long-term spaceflight. Scientists will study the data on Misha and me for the rest of our lives and beyond. Our space agencies won't be able to push out farther into space, to a destination like Mars, until we can learn more about how to strengthen the weakest links in the chain that makes spaceflight possible: the human body and mind. People often ask me why I volunteered for this mission, knowing the risks—the risk of launch, the risk inherent in spacewalks, the risk of returning to Earth, the risk I would be exposed to every moment I lived in a metal container orbiting the Earth at 17,500 miles per hour. I have a few answers I give to this question, but none of them feels fully satisfying to me. None of them quite answers it.

When I was a boy, I had a strange recurring daydream. I saw myself confined to a small space, barely big enough to lie down in. Curled up on the floor, I knew that I would be there for a long time. I couldn't leave, but I didn't mind—I had the feeling I had everything I needed. Something about that small space, the sense that I was doing something challenging just by living there, was appealing to me. I felt I was where I belonged.

One night when I was five, my parents shook Mark and me awake and hustled us down to the living room to watch a blurry gray image on TV, which they explained was men walking on the moon. I remember hearing the staticky voice of Neil Armstrong and trying to make sense of the outrageous claim that he was visiting the glowing disc in the New Jersey summer sky I could see out our window. Watching the moon landing left me with a strange recurring nightmare: I dreamed I was preparing to launch on a rocket to the moon, but rather than being secured safely in a seat inside, I was instead strapped across the pointy end of the rocket, my back against its nose cone, facing straight up at the heavens. The moon loomed over me, its giant craters threatening, as I waited through the countdown. I knew I couldn't possibly survive the moment of ignition. Every time I had this dream, I woke up, sweating and terrified, just before the engines burned their fire into the sky.

As a kid, I took all the risks I could, not because I was foolhardy but because everything else was boring. I threw myself off things, crawled under things, took dares from other boys, skated and slid and swam and capsized, sometimes tempting death. Mark and I climbed up drainpipes starting when we were six, waving back down at our parents from roofs two or three stories up. Attempting something difficult was the only way to live. If you were doing something safe, something you already knew could be done, you were wasting time. I found it bewildering that some people my age could just sit still, breathing and blinking, for entire school days—that they could resist the urge to run outside, to take off exploring, to do something new, to take risks. What went through their heads? What could they learn in a classroom that could even approach the feeling of flying down a hill out of control on a bike?

I was a terrible student, always staring out windows or looking at the clock, waiting for class to be over. My teachers scolded, then chastised, then finally—some of them—ignored me. My parents, a cop and a secretary, tried unsuccessfully to discipline my brother and me. Neither of us listened. We were on our own much of the time—after school, while our parents were still at work, and on weekend mornings, when our parents were sleeping off a hangover. We were free to do what we liked, and what we liked was to take risks.

During my high school years, for the first time I found something I was good at that adults approved of: I worked as an emergency medical technician. When I took the EMT classes, I discovered that I had the patience to sit down and study. I started as a volunteer and in a few years worked my way up to a full-time job. I rode in an ambulance all night, never knowing what I would face next—gunshot wounds, heart attacks, broken bones. Once I delivered a baby in a public housing project, the mother in a rancid bed with old unwashed sheets, a single naked lightbulb swinging overhead, dirty dishes piled in the sink. The heart-pounding feeling of walking into a potentially dangerous situation and having to depend on my wits was intoxicating. I was dealing with life-and-death situations, not boring—and, to me, pointless—classroom subjects. In the morning, I often drove home and went to sleep instead of going to school.

I managed to graduate from high school, in the bottom half of my class. I went to the only college I was accepted to (which was a different college than the one I had meant to apply to—such were my powers of concentration). There, I had no more interest in schoolwork than I'd had in high school, and I was also getting too old to jump off things for fun. Partying took the place of physical risk, but it wasn't as satisfying. When asked by adults, I said I wanted to be a doctor. I'd signed up for premed classes but was failing them in my first semester. I knew I was just marking time until I'd be told I would have to do something else, and I had no idea what that would be.

One day I walked into the campus bookstore to buy snacks, and a display caught my eye. The letters on the book's cover seemed to streak into the future with unstoppable speed: *The Right Stuff.* I wasn't much of a reader—whenever I was assigned to read a book for school, I would barely flip through it, hopelessly bored. Sometimes I'd look at the CliffsNotes and remember enough of what I read to pass a test on the book, sometimes not. I had not read many books by choice in my entire life—but this book somehow drew me to it.

I picked up a copy, and its first sentences dropped me into the stench of a smoky field at the naval air station in Jacksonville, Florida, where a young test pilot had just been killed and burned beyond recognition. He had crashed his airplane into a tree, which "knocked [his] head to pieces like a melon." The scene captured my attention like nothing else I had ever read. Something about this was deeply familiar, though I couldn't say what.

I bought the book and lay on my unmade dorm room bed reading it for the rest of the day, heart pounding, Tom Wolfe's hyperactive, looping sentences ringing in my head. I was captivated by the description of the Navy test pilots, young hotshots catapulting off aircraft carriers, testing unstable airplanes, drinking hard, and generally moving through the world like exceptional badasses.

> The idea here (in the all-enclosing fraternity) seemed to be that a man should have the ability to go up in a hurtling piece of machinery and put his hide on the line and then have the moxie, the reflexes, the experience, the coolness, to pull it back in the last yawning moment—and then to go up again *the next day,* and the next day, and every next day, even if the series should prove infinite—and, ultimately, in its best expression, do so in a cause that means something to thousands, to a people, a nation, to humanity, to God.

This wasn't just an exciting adventure story. This was something more like a life plan. These young men, flying jets in the Navy, did a real job that existed in the real world. Some of them became astronauts, and that was a real job too. These were hard jobs to get, I understood, but some people did get them. It could be done. What drew me to these Navy pilots wasn't the idea of the "right stuff"—a special quality these few brave men had—it was the idea of doing something immensely difficult, risking your life for it, and surviving. It was like a night run in the ambulance, but at the speed of sound. The adults around me who encouraged me to become a doctor thought I liked being an EMT because I liked taking people's blood pressure measurements, stabilizing broken bones, and helping people. But what I craved about the ambulance was the excitement, the difficulty, the unknown, the risk. Here, in a book, I found something I'd thought I would never find: an ambition. I closed the book late that night a different person.

I would be asked many times over the following decades what the beginning of my career as an astronaut was, and I would talk about seeing the moon landing as a kid, or seeing the first shuttle launch. These answers were to some extent true. I never told the story about an eighteen-year-old boy in a tiny, stuffy dorm room, enthralled by swirling sentences describing long-dead pilots. That was the real beginning.

WHEN I BECAME an astronaut and started getting to know my astronaut classmates, many of us shared the same memory of coming downstairs in our pajamas as little kids to watch the moon landing. Most of them had decided, then and there, to go to space one day. At the time, we were promised that Americans would land on the surface of Mars by 1975, when I was eleven. Everything was possible now that we had put a man on the moon. Then NASA lost most of its funding, and our dreams of space were downgraded over the decades. Yet my astronaut class was told we would be the first to go to Mars, and we believed it so fully that we put it on the class patch we wore on our flight jackets, a little red planet rising above the moon and the Earth. Since then, NASA has accomplished the assembly of the International Space Station, the hardest thing human beings have ever achieved. Getting to Mars and back will be even harder, and I have spent a year in space—longer than it would take to get to Mars—to help answer some of the questions about how we can survive that journey.

The risk taking of my youth is still with me. My childhood memories are of the uncontrollable forces of physics, the dream of climbing higher, the danger of gravity. For an astronaut, those memories are unsettling in one way but comforting in another. Every time I took a risk, I lived to draw breath again. Every time I got myself into trouble, I made it out alive.

Most of the way through my yearlong mission, I was thinking about how much *The Right Stuff* had meant to me, and I decided to call Tom Wolfe; I thought he might enjoy getting a call from space. Among the other things we talked about, I asked him how he writes his books, how I might start to think about putting my experiences into words.

"Begin at the beginning," he said, and so I will.

1

February 20, 2015

You have to go to the ends of the Earth in order to leave the Earth. Since the space shuttles were retired in 2011, we've depended on the Russians to launch us into space, and we must start with a journey to the Baikonur Cosmodrome on the desert steppes of Kazakhstan. First, I fly from Houston to Moscow, a familiar journey of eleven hours, and from there ride in a van to Star City, Russia, forty-five miles away—anywhere from one to four hours, depending on Moscow traffic. Star City is the Russian equivalent of the Johnson Space Center; it's the place where the cosmonauts have been trained for the last fifty years (and, more recently, the astronauts who will travel to space with them).

Star City is a town with its own mayor and a church, museums, and apartment blocks. There is a giant statue of Yuri Gagarin, who became the first human in space in 1961, taking a simple, humble socialist-realist step forward while holding a bouquet of flowers behind his back. Years ago, the Russian space agency built a row of town houses especially for us Americans, and staying in them is sort of like staying on a movie set based on a Russian stereotype of how Americans live. There are huge fridges and huge TVs but somehow everything is slightly off. I've spent a lot of time in Star City, including serving as NASA's director of operations there, but it still feels foreign to me, especially in the heart of the frozen Russian winter. After a few weeks of training, I find myself longing to head back to Houston.

From Star City we fly 1,600 miles to Baikonur, once the secret launch site for the Soviet space program. People sometimes say that a place is "in the middle of nowhere," but I never say that anymore unless I'm talking about Baikonur. The launch site was actually built in a village called Tyuratam, named for a descendant of Genghis Khan, but was referred to as Baikonur, the name of another town several hundred miles away, as subterfuge. Now this is the only place called Baikonur. Early on, the Soviets also referred to their launch facility as Star City so as to further confuse the United States. For an American who grew up and trained as a Navy pilot during the tail end of the Cold War, it will always feel a bit strange that I'm invited into the epicenter of the former Soviet space program to be taught its secrets. The people who live in Baikonur now are mostly Kazakh, descendants of Turkic and Mongol tribes, with a minority of ethnic Russians who were left behind after the breakup of the Soviet Union. Russia leases the facilities here from Kazakhstan. The Russian ruble is the main currency, and all the vehicles have Russian license plates.

From the air, Baikonur seems to have been flung randomly onto the high desert steppes. It's a strange collection of ugly concrete buildings, horribly hot in summer and harshly cold in winter, with mounds of rusting, disused machinery piled everywhere. Packs of wild dogs and camels scrounge in the shadows of aerospace equipment. It's a desolate and brutal place, and it's the only working human spaceport for most of the world.

I'm descending toward Baikonur in a Tupolev 134, an old Russian military transport plane. This aircraft might once have been outfitted with bomb racks and in a pinch could have served as a bomber, part of the Cold War arsenal the Soviets developed with the purpose of attacking my country. But now it's used to transport international crews of space travelers—Russian, American, European, Japanese, and Canadian. We are former enemies remade as crewmates, on our way to the space station we built together.

The front of the plane is reserved for the prime crew (my two Russian crewmates and me) and a number of VIPs. Occasionally I wander toward the back, where I've flown on previous trips to Baikonur. Everyone has been drinking since we left Star City this morning, and the junior Russian personnel have their own party going on back here. Russians never drink without eating, so in addition to vodka and cognac they are serving tomatoes, cheese, sausage, pickled cucumbers, dried salty fish chips, and slices of salted pork fat called *salo*. On my first trip to Kazakhstan, in 2000, I was making my way through the party at the back of the plane to find the bathroom when I was stopped and forced to drink shots of *samogon*, Russian moonshine. The junior guys were so drunk they were stumbling around from the turbulence and alcohol, spilling the stuff on themselves and on the floor of the plane, all while chain-smoking. We were lucky to have made it to Kazakhstan without exploding into a giant fireball of moonshine and jet fuel.

Today everyone is drinking heavily again, and we are pretty well fueled by the time we plunge from the clouds over the flat, frozen desert and touch down on Baikonur's single runway. We climb out, blinking in the cold, and encounter a welcoming party: officials from Roscosmos, the Russian space agency, and Energia, the company that builds the Soyuz spacecraft, one of which will take us into orbit to dock with the International Space Station. The mayor of Baikonur is here, as well as other local dignitaries. My Russian crewmate Gennady Padalka strides forward and speaks sternly to them as we stand at semi-attention, "*My gotovy k sleduyushchim shagam nashey podgotovki.*" ("We are ready for the next steps of our preparation.")

This is a ritual, like so many in spaceflight. We Americans have similar staged moments at similar points of launch preparation. There is a fine line between ritual and superstition, and in a life-threatening business such as spaceflight, superstition can be comforting even to the nonbeliever.

We see a strange but welcoming sight at the edge of the tarmac: a group of Kazakh kids, little ambassadors from the end of the Earth. They are round cheeked, black haired, mostly Asian in appearance, wearing bright, dusty clothes and holding balloons. The Russian flight doctor has warned us to stay away from them: there has been concern about a measles outbreak in this region, and if one of us were to be infected it would bring serious consequences. We have all been vaccinated, but the Russian flight surgeons are very cautious; no one wants to go to space with measles. Normally we do what the doctor says, especially since he has the power to ground us. But Gennady walks confidently forward anyway.

"We must say hello to the children," he says firmly in English.

I've known both Gennady and our third crewmate, Mikhail Kornienko ("Misha"), since the late 1990s, when I started traveling to Russia to work on the joint space station program between our two countries. Gennady has a thick head of silver hair and a sharp gaze that doesn't miss much. He is fifty-six and is the commander of our Soyuz. He's a natural leader, gruffly shouting out orders when necessary but listening carefully when one of his crew has another perspective. He's a person I trust implicitly. Once, in Moscow, near the Kremlin, I saw him break away from his fellow cosmonauts to pay his respects at the site where an opposition politician had been murdered, possibly by surrogates of Vladimir Putin. For a cosmonaut, an employee of Putin's government, that gesture was risky. The other Russians with us seemed to be reluctant even to discuss the murder, but not Gennady.

Misha, who will be my fellow traveler for a year, is fifty-four and is very different from Gennady—casual, quiet, and contemplative. Misha's father was a military helicopter pilot working with the cosmonaut rescue forces, and when Misha was only five, his father died in a helicopter crash. His early dreams of flying in space were only reinforced by this unfathomable loss. After serving in the military as a paratrooper, Misha needed to get a degree in engineering from the Moscow Aviation Institute to qualify as a flight engineer. He couldn't get in, because he wasn't a resident of the Moscow region, so he became a Moscow police officer to establish residency and was then able to study at the institute. He was selected as a cosmonaut in 1998.

When Misha stares at you with his light blue eyes, it feels like nothing is more important to him than fully comprehending what you are saying. He is more open with his feelings than the other Russians I know. If he were American, I could picture him as a Birkenstock-wearing hippie living in Vermont.

We approach the Kazakh kids gathered to welcome us. We greet them, shake hands, and accept flowers that for all I know are covered with measles. Gennady chats with the children happily, his face lit up with his signature smile.

The entire party—prime crew, backup crew, and support staff—boards two buses for the ride to the quarantine facility where we will spend the next two weeks. (The prime crew and backup crew always travel separately, for the same reason the president and vice president do.) As we are boarding, for a laugh, Gennady sits in the driver's seat of our bus, and we all take pictures of him with our phones. Many years ago, crews used to travel to Baikonur, spend one day here checking out the Soyuz spacecraft, then travel back to Star City to wait out the two weeks until launch. Now, cutbacks require that we make only one trip, so we will be stuck here for the duration. I take a window seat, pop in my earbuds, and rest my head against the window, hoping to become sleepy enough to take a nap before we get to our hotel-like quarantine facility. This road is in terrible shape—it always has been, and it only gets worse—and the rutted and patched asphalt rattles my head against the window enough to keep me awake.

We pass dilapidated Soviet-era apartment complexes, huge rusted satellite dishes communicating with Russian spacecraft, mounds of garbage randomly strewn about, the occasional camel. It's a clear, sunny day. We pass Baikonur's own statue of Yuri Gagarin, this one with his arms raised—not in the triumphant V of a gymnast celebrating a perfect dismount, but the joyful straight-up gesture of a kid about to try a somersault. In this statue, he's smiling.

Far over the horizon a launch tower rises above the same deteriorating concrete pad from which Yuri first rocketed off Earth, the same pad from which nearly every Russian cosmonaut has left Earth, the same pad from which I will leave Earth two weeks from now. The Russians sometimes seem to care more about tradition than they do about appearance or function. This launchpad, which they call Gagarinsky Start (Gagarin's Launchpad), is imbued with the successes of the past, and they have no plans to replace it.

Misha's and my mission to spend a year on ISS is unprecedented. A normal mission to the space station lasts five to six months, so scientists have a good deal of data about what happens to the human body in space for that length of time. But very little is known about what happens after month six. The symptoms might get precipitously worse in the ninth month, for instance, or they might level off. We don't know, and there is only one way to find out.

Misha and I will collect various types of data for studies on ourselves, which will take a significant amount of our time. Because Mark and I are identical twins, I'm also taking part in an extensive study comparing the two of us throughout the year, down to the genetic level. The International Space Station is a world-class orbiting laboratory, and in addition to the human studies of which I am one of the main subjects, I will also spend a lot of my time this year working on other experiments, like fluid physics, botany, combustion, and Earth observation.

When I talk about the International Space Station to audiences, I always share with them the importance of the science being done there. But to me, it's just as important that the station is serving as a foothold for our species in space. From there, we can learn more about how to push out farther into the cosmos. The costs are high, as are the risks.

On my last flight to the space station, a mission of 159 days, I lost bone mass, my muscles atrophied, and my blood redistributed itself in my body, which strained and shrank the walls of my heart. More troubling, I experienced problems with my vision, as many other astronauts have. I have been exposed to more than thirty times the radiation of a person on Earth, equivalent to about ten chest X-rays every day. This exposure will increase my risk of a fatal cancer for the rest of my life. None of this compares, though, to the most troubling risk: that something bad could happen to someone I love while I'm in space with no way for me to come home.

Looking out the window at the strange landscape of Baikonur, I realize that for all the time I've spent here, weeks in fact, I have never really seen the town itself. I've only been to the designated spaces where I have official business: the hangars where the engineers and technicians prepare our spacecraft and rocket for flight; the windowless fluorescent-lit rooms where we get into our Sokol pressure suits; the building where our instructors, interpreters, doctors, cooks, management people, and other support staff stay; and the nearby building, affectionately called Saddam's Palace by Americans, where we stay. This opulent residence was built to accommodate the head of the Russian space agency and his staff and guests, and he allows the crews to use it while we're here. It's a nicer place than the other facility, and far nicer than the austere crew quarters housed in an office building where shuttle astronauts used to spend quarantine at the Kennedy Space Center in Florida. Saddam's Palace has crystal chandeliers, marble floors, and a four-room suite for each of us, complete with Jacuzzi tubs. The building also has a *banya,* or Russian sauna, with a cold pool to jump into afterward. Early in our two-week quarantine I go out to the *banya* to find a naked Misha beating on a naked Gennady with birch branches. The first time I saw this scene I was a bit taken aback, but once I experienced the *banya* myself, followed by a dip in a freezing cold pool of water and a homemade Russian beer, I completely understood the appeal.

Saddam's Palace also has an elaborate dining room with pressed white tablecloths, fine china, and a flat-screen TV on the wall that constantly plays old Russian movies the cosmonauts seem to love. The Russian food is good, but for Americans it can start to get old after a while—borscht at nearly every meal, meat and potatoes, other kinds of meat and potatoes, everything covered with tons of dill.

"Gennady," I say while we are eating dinner a few days into our stay. "What's with all the dill?"

"What do you mean?" he asks.

"You guys put dill on everything. Some of this food might actually be good if it weren't covered in dill."

"Ah, okay, I understand," Gennady says, nodding, his signature smile starting to emerge. "It's from when the Russian diet consisted mostly of potatoes, cabbage, and vodka. Dill gets rid of farts."

Later I google it. It's true. As it happens, getting rid of gas is a worthwhile goal before being sealed into a small tin can together for many hours, so I stop complaining about the dill.

. . .

THE DAY AFTER we arrive in Baikonur we have the first "fit check." This is our opportunity to get inside the Soyuz capsule while it's still in the hangar and not yet attached to the rocket that will launch us into space. In the cavernous hangar known as Building 254 we pull on our Sokol suits—always an awkward process. The only entrance into the suits is in the chest, so we have to slide our lower bodies into the chest hole, then struggle to fit our arms into the sleeves while blindly pulling the neck ring up over our heads. I often come away from this procedure with scratches on my scalp. Not having hair is a disadvantage here. The chest opening is then sealed through a disconcertingly low-tech process of gathering the edges of the fabric together and securing them with elastic bands. The first time I was introduced to this system I couldn't believe those rubber bands were all that was meant to protect us from space. Once I got to the space station, I learned the Russians use the exact same rubber bands to seal their garbage bags in space. In one sense I find this comical; in another way I respect the efficiency of the Russian philosophy on technology. If it works, why change it?

The Sokol suit was designed as a rescue suit, which means that its only function is to save us in case of a fire or depressurization in the Soyuz. It's different from the spacesuit I will wear during spacewalks later in my mission; that suit will be much sturdier and more functional, a little spaceship in its own right. The Sokol suit serves the same purpose as the orange NASA-designed pressure suit I used to wear to fly on the space shuttle. NASA introduced that suit only after the *Challenger* disaster of 1986; before that, shuttle astronauts wore simple cloth flight suits, as the Russians did before a depressurization accident killed three cosmonauts in 1971. Since then, cosmonauts (and any astronauts to join them in the Soyuz) have worn the Sokol suits. In a weird way, we are surrounded by the evidence of tragedies, too-late fixes that might have saved the astronauts and cosmonauts who took the same risks we are taking and lost.

Today is like a dress rehearsal: we suit up, perform leak checks, then get strapped into our custom-made seats, built from plaster molds taken of our bodies. This is not for our comfort, which is not a particularly high priority for the Russians, but for safety and to save space—no sense building more seat than is strictly necessary. The custom-molded seats will cradle our spines and absorb some of the impact on the hard return to Earth a year after we depart.

As much time as I have spent in the Soyuz simulators in Star City, I'm still amazed by how difficult it is to wedge myself and my pressure suit into my seat. Each time, I have a moment of doubt whether I'm going to fit. But then I do—just barely. If I were to sit up out of the seat liner, my head would smash against the wall. I wonder how my taller colleagues do it. Once we're strapped in, we practice using the hardware, reaching out for buttons, reading data off screens, grabbing our checklists. We discuss what we might want to have customized for us, down to details such as where we want our timers (used for timing engine burns), where we want our pencils, and where we want the bits of Velcro that will allow us to put things "down" in space.

When we finish, we clamber back out of the hatch and look around the dusty hangar. The next Progress resupply vehicle is here; it looks very similar to the Soyuz, because the Russians never create two designs when one will do. In a few months, this Progress will deliver equipment, experiments, food, oxygen, and care packages to us on the ISS. After that, a Soyuz will launch in July carrying a new three-person crew. Somewhere in this hangar, the parts for the next Soyuz to fly after that are being assembled, and another one after that, and another one after that. The Russians have been launching the Soyuz since I was three years old.

The Soyuz spacecraft—Soyuz means "union," as in "Soviet Union," in Russian—is designed to maneuver in space, dock with the station, and keep human beings alive, but the rockets are the workhorses, humanity's answer to the pull of Earth's gravity. The rockets (which for some odd reason are also called Soyuz) are prepared for launch in an assembly and test facility across from the hangars known as site 112. Gennady, Misha, and I cross the street, pass the gathered Russian media, enter that enormous building, and stand in another cavernous, quiet room, this time regarding our rocket. Gunmetal gray, it lies on its side. Unlike the space shuttle or the colossal Apollo/Saturn that preceded it, the Soyuz spacecraft and rocket combination is assembled horizontally and rolled out to the launchpad in that position. Only when it reaches the launchpad, a couple of days before we launch, will it be set upright to vertical, pointing toward its destination. This is yet another example of how the Russians and Americans do things differently. In this case, the procedure is less ceremonial than the NASA way, with its stately, majestic rollouts of vertical launch vehicles balanced on an enormous crawler transporter.

At 162 feet long, this rocket, the Soyuz-FG, is noticeably smaller than the assembled space shuttle, but it's still a daunting colossus, a building-size object that will, we hope, leave the ground, with us riding on top of it, at twenty-five times the speed of sound. Its navy gray sheet metal, adorned with low-tech rivets, is unbeautiful but somehow comforting in its utility. The Soyuz-FG is the grandchild of the Soviet R-7, the world's first intercontinental ballistic missile. The R-7 was designed during the Cold War for launching nuclear weapons at American targets, and I can't help remembering how as a child I was aware that New York City, and my suburb of West Orange, New Jersey, would certainly have been among the first targets to be instantly vaporized by a Soviet attack. Today, I'm standing inside their formerly secret facility, discussing with two Russians our plans to trust one another with our lives while riding to space on this converted weapon.

Gennady, Misha, and I all served in our militaries before being chosen to fly in space, and though it's something we never talk about, we all know we could have been ordered to kill one another. Now we are taking part in the largest peaceful international collaboration in history. When people ask whether the space station is worth the expense, this is something I always point out. What is it worth to see two former bitter enemies transform their weapons into transport for exploration and the pursuit of scientific knowledge? What is it worth to see former enemy nations turn their warriors into crewmates and lifelong friends? This is impossible to put a dollar figure on, but to me it's one of the things that makes this project worth the expense, even worth risking our lives.

. . .

THE INTERNATIONAL SPACE STATION got its start in 1984, when President Reagan announced during his State of the Union address that NASA was designing a space station, *Freedom*, to be put in orbit within ten years. Resistance from Congress created years of cutbacks and reconfigurations, and *Freedom* was no closer to actually being built when, in 1993, President Clinton announced that the station would be merged with the Russian Federal Space Agency's proposed space station *Mir-2*. With the addition of space agencies representing Europe, Japan, and Canada, the international coalition came to include fifteen countries. It took more than one hundred launches to get the components into orbit and more than one hundred spacewalks to assemble them. The ISS is a remarkable achievement of technology and international cooperation. It has been inhabited nonstop since November 2, 2000; put another way, it has been more than fourteen years since all humans were on the Earth at once. It is by far the longest-inhabited structure in space and has been visited by more than two hundred people from sixteen nations. It's the largest peacetime international project in history.

I wake up my last morning on Earth at around seven. I spend the morning looking through all of my bags—one to meet me in Kazakhstan, others to send back to Houston. The logistics are strange: What do I want to have with me when I first land? What will I not need until later? Did I make sure to write down my credit card and account numbers for things like my utilities and bank? It's hard enough to deal with these sorts of details on Earth, but I need to be prepared to avoid defaulting on my mortgage, as well as to buy Amiko and the girls presents, from space.

My last earthly breakfast is a Baikonur attempt at American cuisine: runny eggs (because I could never make the Kazakh cook understand the term "over medium"), toast, and "breakfast sausages" (actually microwaved hot dogs). Getting ready on the day of launch takes much longer than you'd think it would, like so many aspects of spaceflight. First I take a final trip to the *banya* to relax, then go through the preflight enema ritual—our guts shut down in space initially, so the Russians encourage us to get things cleaned out ahead of time. The cosmonauts have their doctors do this, with warm water and rubber hoses, but I opt for the drugstore type in private, which lets me maintain a comfortable friendship with my flight surgeon. I savor a bath in the Jacuzzi tub, then a nap (because our launch is scheduled for 1:42 a.m. local time). When I wake, I take a shower, lingering awhile. I know how much I'll miss the feeling of water for the next year.

The Russian flight surgeon we call "Dr. No" shows up shortly after I'm out of the shower. He is called Dr. No because he gets to decide whether our families can see us once we're in quarantine. His decisions are arbitrary, sometimes mean-spirited, and absolute. He is here to wipe down our entire bodies with alcohol wipes. The original idea behind the alcohol swab-down was to kill any germs trying to stow away with space travelers, but now it seems like just another ritual. After a champagne toast with senior management and our significant others, we sit in silence for a minute, a Russian tradition before a long trip. As we leave the building, a Russian Orthodox priest will bless us and throw holy water into each of our faces. Every cosmonaut since Yuri Gagarin has gone through each of these steps, so we will go through them, too. I'm not religious, but I always say that when you're getting ready to be rocketed into space, a blessing can't hurt.

We do a ceremonial walkout past the media with a traditional Russian song playing, "Trava u doma," a song about cosmonauts missing home that sounds like a Soviet marching band playing at a carnival:

> *And we don't dream of the cosmodrome's roar*
> *Nor of this icy dark blue*
> *Instead, we dream of the grass, the grass near our homes . . .*
> *The green, green grass.*

We get on the bus that will take us to the building where we get suited up. The moment the door to the bus closes behind us, a rope holding back the crowd is cut, and everyone rushes forward. It's chaotic, and I can't spot my family at first, but then I see them, in the front row—

Amiko, Samantha, Charlotte, and Mark. Someone lifts up Charlotte, who is eleven, so she can put her hand on the window, and I put my hand up to hers, trying to look happy. Charlotte is smiling, her round white face in a grin. If she's sad that she won't see me for a year, if she's scared to watch me leave Earth on a barely controlled bomb, if she's aware of the many types of peril I will face before I get to hug her again a year from now, she doesn't show it. Then she's down on the asphalt again, standing with the rest of them and waving. I see Amiko smiling, though I can also see tears in her eyes. I see Samantha, who is twenty. Her wide smile betrays her apprehension for what is to come. And then the bus's brakes release with a hissing scream, and we are gone.

I SIT ON a cheap leather couch waiting to suit up in Building 254, a thirty-minute drive from Saddam's Palace. A flat-screen in the corner shows a silly Russian TV show that none of us pay attention to. There is some food laid out—cold chicken, meat pies, juice, and tea—and though it isn't what I would have chosen as my last earthly meal for a year, I eat a bit.

First Gennady is called into an adjacent room to strip down and put on his diaper, cardiac electrodes, and a fresh pair of white long underwear (meant to absorb sweat and shield us from the rubber of the Sokol suit). When Gennady returns, Misha goes in to diaper up. Then I go. Whenever I do this, I chuckle to myself that I wouldn't have thought I'd be in diapers again until much later in life. It's now time to get our Sokol suits on. We have white-coated, surgical-masked Russian specialists to help us get dressed. They expertly close the openings in our suits with a series of folds and the peculiar rubber bands.

The three of us walk into another room, which is divided by a sheet of glass. On the other side are our families, managers from the Russian space agency (Roscosmos), NASA leaders, and members of the media, sitting in rows of seats facing us. I know the closest analogy should be a NASA press conference, but this moment has always made me feel like a gorilla in the zoo instead.

I immediately spot Amiko, Mark, and my daughters in the front row. Amiko and the girls have been here for a few days, but Mark only just arrived. They all smile at me and wave. Not for the first time, I'm grateful that my brother is there for them. As an experienced astronaut himself, and someone who knows me better than anyone else, he can help them understand what will be going on, and reassure them when necessary, better than anyone else could.

Amiko smiles happily and points to the pendant I had made for her before I left Houston, a silver version of the Year in Space mission patch. Samantha and Charlotte wear the silver pendants too. I will bring a second version of the pendants, in gold with sapphires set into them, with me to give the three of them when I return. Amiko's smile is sincere and happy, but because I know her so well I can also see that she is tired, not only from jet lag but also from the strain. This is Amiko's second time going through the process of preparing for a long-duration mission with me, so she's known what to expect, though I'm not sure that makes it any easier. She also works for NASA, in the public affairs office, so she knows better than most astronauts' partners what I am facing on this mission. In some instances this knowledge will be comforting, but more often than not—including today—it would have been less stressful to know less.

Amiko was an acquaintance for a long time. She'd worked closely with my brother on a project and also shared mutual friends with my ex-wife, Leslie. In early 2009, Amiko and I both filed for divorce, each unbeknownst to the other, and by coincidence we ran into each other a few times many months later. Amiko says she remembers one of those nights being impressed that though I had jokingly acknowledged she was attractive, I'd turned down a chance to get in the hot tub with her and some other people in favor of going to bed early for a training event in the morning. A few weeks later, I saw her again at a party, and this time I did wind up in the hot tub with her. We talked all night, but again she was impressed that I didn't make any advances. Anyone who has seen Amiko knows that she has to deal with a lot of attention from men, and I guess I stood out for trying to get to know her as a person. But I'm not an idiot—I did make sure to get her phone number that night.

I'm always curious about how people end up doing what they do, especially when they seem to be especially good at it, as Amiko is. She struck me as different from a lot of the people who work for NASA public affairs, some of whom can be conservative about new ideas and resistant to change. I asked her about how she came to her career, and though she told me only a bit of her history it was pretty compelling. Kicked out of her house at fifteen for standing up to her mother's physical abuse, with nothing but the clothes on her back, married at eighteen, and with two kids by twenty-three, she got a job as a NASA secretary. From the moment she was hired, she began working toward getting into the agency's competitive employee education program, in which NASA pays for promising employees to take college classes. Once Amiko was chosen, she began taking the maximum credits possible every semester while working full-time and raising her two small boys. She finished her degree in communications with a perfect GPA and every honor given to undergraduates. I had known she was smart and capable, but the more I learned about her life story, the more impressed I was with her. Her two sons, by that point in high school, were doing well, and she continued to set new challenges for herself. Most people would not have gotten past the setbacks she'd faced, but through intelligence, grit, and fierce determination, she had made the life for herself that she wanted. I could tell she wouldn't easily change that life for any man, even for an astronaut being as charming as he could.

We started seeing each other that fall, and things between us had become serious by the time I went to space in October 2010. This was my first long-duration mission to the International Space Station and her first mission as the partner left behind. It was an unusual challenge for a new relationship. We were both surprised to find that the separation only brought us closer. I could depend on her as my partner on the ground, and we enjoyed being able to give each other our undivided attention for the hour or so a day we could talk on the phone. I came back more confident than ever that we belonged together. I know some of our friends wonder why we haven't gotten married—we've been together now for five and a half years and have lived together for much of that time. I've been there for her sons when necessary, and she is always there for my daughters. We are as committed as any married couple, but because both of us have been married before, and neither of us is especially traditional, we don't seem to see the point in it. The media sometimes refer to Amiko as my "longtime partner," and that seems right to both of us.

Sitting next to Amiko is Samantha. I'd been surprised to see her new look when she showed up in Baikonur, her long curls dyed black, thick black eyeliner, dark red lipstick, and nothing but black clothes. Since her mother and I divorced, my relationship with Samantha has been rocky, and in many ways it's still recovering from the fallout. She was fifteen when Leslie moved the girls against my wishes from Houston to Virginia Beach, an especially tough age to deal with that kind of upheaval. Samantha blames me for the divorce and for many of the problems that have come since. When I look at her today through the glass, her blue eyes sparkling under the heavy eyeliner, I still see her the way she looked the first time I saw her, in 1994, in the maternity ward of the Patuxent River naval air station, where I was a test pilot. Leslie went through a long, difficult labor, and when Samantha was finally delivered it was by emergency cesarean section. When I first saw her tiny pink face with one eye shut and the other eye open, I felt an unbelievable urge to protect her. Though she's an adult now, I still feel the same way.

Charlotte was born when Samantha was almost nine, an age gap that has made it easy for them to get along. Samantha seems to enjoy having an adoring sidekick, and Charlotte has had the freedom to go anywhere her older sister is willing to take her—including to Baikonur. Charlotte's birth was even more difficult than Samantha's; I remember standing in the operating room and hearing the doctor calling an emergency code. When they finally got Charlotte out, she was limp and unresponsive. I still remember the sight of her tiny, blue, lifeless arm hanging out of the incision. The doctors warned us she might have cerebral palsy, but she has grown up to be healthy, bright, strong, and a generous-spirited person. I know she must be experiencing extremes of emotion today, but she seems happy and calm, sitting next to her sister and brushing her light brown bangs out of her eyes to smile at me. I feel grateful that my daughters are able to lean on Amiko for reassurance and to follow her lead in how to deal with the stresses of this week.

I also spot Spanky—Mike Fincke, a friend and colleague from my astronaut class—who has been in charge of helping out my family while I've been in quarantine. Between missions, astronauts can be assigned to take on all kinds of earthbound responsibilities, and Spanky, who has been to ISS himself and will probably go back, has been fantastic with my family—answering questions, fulfilling special requests, communicating their preferences to NASA whenever possible. This is the second time Spanky has done this job for me.

On our side of the glass is a mock-up of the Soyuz seat, and one by one, Gennady, Misha, and I get into it, lying on our backs. Technicians check our suits for leaks. I lie there for fifteen minutes with my helmet closed and my knees pressed up to my chest while a large room full of people, some of whom I don't know, watch politely. Why we need to do this for an audience I've never been sure—another ritual. Afterward we sit in a row of chairs before the glass to have a last talk, through microphones, with our families.

The things we want to say to our loved ones before we might be about to die in a fireball above Kazakhstan are not the things we would want to say while the assembled press from a number of countries listen from rows of chairs and write down our every word. Adding to the awkwardness, we are all sharing one audio system, so each family has to wait their turn to avoid talking over one another. Still, I wouldn't want my daughters' last image to be of me speaking a few terse words into a microphone, so I try to split the difference by saying little but trying to communicate much in other ways, figuring that simple gestures can say a lot. I give Amiko and the girls the "I've got my eyes on you" gesture, pointing back and forth from my eyes to their eyes. It makes them smile.

When we finish this ritual and go outside, it's dark and freezing cold. Floodlights blind us as we walk into the parking lot, flanked by rows of media and spectators we can barely make out. The Sokol suits are designed for sitting in the fetal position while launching in the

Soyuz, not for walking, so the three of us waddle along like hunched penguins with as much dignity as we can. We are carrying cooling fans that blow air into our pressure suits, like the Apollo astronauts in the old NASA footage. We are all wearing two pairs of thin white gloves that are meant to keep us from bringing germs to space (at least, that's the idea). We will remove the top layer right before we get into the Soyuz.

The bus taking us to the launchpad is idling nearby, its billowing exhaust silhouetted by the floodlights. The three of us walk up to three small white squares that have been painted on the asphalt, labeled with our positions on the Soyuz: commander for Gennady, flight engineer for Misha, flight engineer 2 for me. We step into our little boxes and wait for the head of the Russian space agency to ask us each in turn, again, if we are ready for our flight. It's sort of like getting married, except whenever you're asked a question you say, "We are ready for the flight" instead of "I do." I'm sure the American rituals would seem just as alien to the Russians: before flying on the space shuttle, we would get suited up in our orange launch-and-entry suits, stand around a table in the Operations and Checkout Building, and then play a very specific version of lowball poker. We couldn't go out to the launchpad until the commander had lost a round (by getting the highest hand), using up his or her bad luck for the day. No one remembers exactly how this tradition got started. Probably some crew did it first and came back alive, so everyone else had to do it too.

We board the bus—the prime crew, our flight surgeons, the Gagarin Cosmonaut Training Center managers, and a few suit technicians. We sit on the side facing all the lights and clamoring people. I catch sight of my family one last time and give them a wave. The bus slowly pulls away, and they are gone.

Soon, we are moving, the motion lulling us into a contemplative trance. After a while, the bus slows, then comes to a stop well before the launchpad. We nod at one another, step off, and take up our positions. We've all undone the rubber-band seals that had been so carefully and publicly leak-checked just an hour before. I center myself in front of the right rear tire and reach into my Sokol suit. I don't really have to pee, but it's a tradition: When Yuri Gagarin was on his way to the launchpad for his historic first spaceflight, he asked to pull over—right about here—and peed on the right rear tire of the bus. Then he went to space and came back alive. So now we all must do the same. The tradition is so well respected that women space travelers bring a bottle of urine or water to splash on the tire rather than getting entirely out of their suits.

This ritual satisfactorily observed, we get back into the bus and resume the last leg of our journey. A few minutes later, the bus makes another stop to let the train pass that has just fueled our rocket. The bus door opens and an unexpected face appears: my brother.

This is a breach of quarantine: my brother, having been on a series of germ-infested planes from the United States to Moscow to Baikonur just yesterday, could be carrying all manner of terrible illnesses. Dr. No has been saying "*Nyet*" all week, and now, suddenly, he sees my brother and says "*Da.*" The Russians enforce the quarantine with an iron fist, then let my brother break it for sentimental reasons; they make a ritual of sealing up our suits, then let us open them to pee on a tire. At times, their inconsistencies drive me nuts, but this gesture, letting me see my brother again when I least expect to, means the world to me. Mark and I don't exchange many words as we ride together for the few minutes out to the launchpad. Here we are, two boys from blue-collar New Jersey who somehow made it such a long way from home.

2

MY EARLIEST MEMORIES ARE of the warm summer nights when my mother tried to settle Mark and me to sleep in our house on Mitchell Street in West Orange, New Jersey. It would still be light outside, and with the windows open the smell of honeysuckle drifted in along with the sounds of the neighborhood—older kids yelling, the thumps of basketballs against driveways, the rustling of breezes high in the trees, the faraway sound of traffic. I remember the feeling of drifting weightless between summer and sleep.

My brother and I were born in 1964. Members of our extended family on my father's side lived all up and down our block, aunts and uncles and cousins. The town was separated by a hill. The more well-off lived "up the hill," and we lived "down the hill," though we wouldn't know until later what that meant in socioeconomic terms. I remember waking early in the morning with my brother when we were small, maybe two years old. My parents were sleeping, so we were on our own. We got bored, figured out how to open the back door, and left the house to explore, two toddlers wandering the neighborhood. We made our way to a gas station, where we played in the grease until the owner found us. He knew where we belonged and stuck us back in the house without waking my parents. When my mother finally got up and came downstairs, she was perplexed by the grease all over us. Later, the owner came over and told her what had happened.

One afternoon when we were kindergarten students, my mother bent down to tell us she had an important responsibility for us. She held a white envelope in front of her as if it were a special prize. She said that we were to put the letter in a mailbox directly across the street from our house. She explained that because it wasn't safe to cross in the middle of the street—we could be hit by a car—we were to walk up to the corner, cross the street there, walk back in this direction on the other side of the street, mail the letter, then retrace our steps all the way back home. We assured her we understood. We walked up to the corner, looked both ways, and crossed. We walked back toward our house on the mailbox side of the street, Mark boosted me up to pull down the heavy blue handle, and I proudly deposited the letter in the slot. Then we pondered our return trip.

"I'm not walking all the way back to the corner," Mark announced. "I'm just going to cross the street right here."

"Mom said we should cross at the corner," I reminded him. "You're going to get hit by a car."

But Mark had made up his mind.

I set off back toward the corner myself, satisfied that I would be praised for having followed directions. (It occurs to me now that following directions that seemed arbitrary was good early training for being an astronaut.) I got to the corner, crossed, and turned back toward the house. The next thing I heard was car brakes squealing and the thump of a collision, and then, out of the corner of my eye, I saw something the size and shape of a kid flying up into the air. The next moment, Mark sat, dazed, in the middle of the street, while the frantic driver fussed over him. Someone ran for our mother, an ambulance came and took them to the hospital, and I spent the rest of the afternoon and evening with my uncle Joe, pondering the different choices Mark and I had made and the different results.

As our childhoods went on, we continued to take crazy risks. We both got hurt. We both got stitches so often we sometimes would have the stitches from the previous injury removed during the same visit new stitches were put in, but only Mark was ever admitted as an inpatient. I was always jealous of the attention he got when he was hospitalized. Mark got hit by the car, Mark broke his arm sliding down a handrail, Mark had appendicitis, Mark stepped on a broken glass bottle of worms and got blood poisoning, Mark was taken into the city for a series of tests to see whether he had bone cancer (he didn't). We both played with BB guns recklessly, but only Mark got shot in the foot and then damaged by a botched surgery.

When we were about five, my parents bought a little vacation bungalow on the Jersey Shore, and some of my best memories from childhood are from that time. It wasn't much more than a shack, with no heat, but we loved going there. My parents would get us up in the middle of the night, when my father got off work, and load us into the back of the family station wagon in our pajamas with our blankets, where we'd go back to sleep. I remember the feeling of being rocked by the car's movement, looking at the telephone wires out the windows and the stars beyond them.

At the shore, in the mornings, Mark and I would ride our bikes to a place called Whitey's, a boatyard where we bought bait for crabbing. We'd spend all day on the dock behind our bungalow, waiting to feel a crab nibble on the bait. We built rafts out of spare fence planks, on which we set sail from the lagoon house on the approach to Barnegat Bay. We had a kind of freedom my own children never had. I remember falling off the dock before I knew how to swim and sinking into the dark and murky water of the lagoon. I didn't know what to do about it. I simply watched the bubbles of the last of my air rising. Then my father, who had seen my blond hair drifting just above the water, grabbed a handful and pulled me out.

MY FATHER WAS an alcoholic, and sometimes he would take off drinking for long periods of time. I remember one weekend at the Jersey Shore when he disappeared, leaving the three of us with no food and no money. My mother explained to us that he had taken our only car to a bar; somehow we got a ride over there to find him. It was a ramshackle place, set off in the marshes that lined Barnegat Bay, built of brown pressure-treated wood that had been bleached by the salt air. He refused to give us any money or to leave with us. I remember my mother's face as she led us out of there. She was upset, but her face showed determination: she would get us through this. We didn't eat that weekend, and I'll never forget how that felt; it affects me to this day when I hear of people who don't have enough to eat. The physical feeling of hunger is horrible, but much worse is the bottomlessness of not knowing when it will end.

When Mark and I were in second grade, our parents sold the place on the Jersey Shore so they could buy a house "up the hill." They wanted us to be able to go to a better public school. We moved onto a street lined with giant green oak trees, aptly named Greenwood Avenue. I remember the smell of springtime on that street, trees with new leaves and azalea bushes of pinks and purples. It's odd that once we moved, we hardly ever saw our family on Mitchell Street again. My father was often not on speaking terms with various friends and family members, so it's possible he had burned through all those relationships by the time we moved.

We may have lived up the hill now, but in socioeconomic terms we still belonged down the hill, sort of like the Beverly Hillbillies we saw on television. We stuck out among the wealthier Jewish families who lived nearby. Mark and I used to get into scrapes with neighbor kids—snowball fights, rock fights, apple fights with the crabapples that fell off the trees. We threw them at adult neighbors too, and we discovered that the grown man next door had a pretty good arm when he threw them back. We were like juvenile delinquents who never got arrested, probably because we were the children of a cop.

In the summertime, my father and his cop buddies would have cookouts in a nearby park, and those days were always fun—at least, at first—as we ate hot dogs and played softball. But as the day went on and the empty bottles and cans piled up, you'd have twenty drunk cops getting into arguments, things turning nasty. My father would finally load us into the car, blind drunk. As he went careening down Pleasant Valley Way, swerving into the opposite lane, we'd be screaming at him not to crash the car.

Sometimes my father's cop friends would come over to our house for parties, and when they got drunk they would pull their guns out. Once, my father wanted to show off his new gun to his partner, so they decided to use a wooden sculpture I had just made in school as a target. I had brought it home and showed it proudly to my parents, and I was heartbroken that my dad would blast holes in my artwork.

My brother and I used to spend one night a week with our paternal grandparents, whom I loved, so my parents could go out drinking. My grandmother, Helen, was a heavy woman who was always impeccably dressed and always wore a wig. She was so pleased to see us every weekend, consistently kind and loving. She let us watch all the TV we wanted and sang us to sleep. My grandfather had served in the Navy in World War II on a destroyer in the Pacific, and it seemed odd to me that after having that kind of extraordinary experience he came home and worked in a mattress factory for the rest of his life. But he was content, had a great sense of humor, and he made a good life for himself and his family despite having only a sixth-grade education. In the mornings our grandparents always took us to the same diner for breakfast. After that we would spend hours visiting the flower gardens surrounding the historic mansions in northern New Jersey. That's how I started to gain an appreciation for flowers, which would come back to me during my year in space, when I was tasked with bringing a crop of zinnias back from the brink of death. As much as I loved the breakfast and the flowers, I loved the routine, the way we did the same things in the same order, the stability of life with my grandparents.

When my brother and I were maybe nine or ten, my parents thought we didn't need to be taken care of anymore when they went out drinking. They'd come home in the middle of the night, drunk and fighting. Kids sleep pretty hard, so the sound would first sneak into my dreams—the shouting and banging starting off low, maybe imaginary. But then it would gradually get louder, and Mark and I would eventually be lying awake, blinking in the dark, hearts pounding, listening to the yelling and screaming and things shattering against the walls.

Sometimes my mother would get scared enough of my father to leave the house with Mark and me. We'd run to my grandparents' house several miles away. We'd bang on the door and wake them in the middle of the night, ask them to take us in. We always wound up going back to our house the next day. I remember coming home on those mornings, feeling like maybe it had all been a dream, but then seeing the things that had been broken scattered on the floor. Sometimes my brother and I would dedicate ourselves to fixing things—plates, furniture, knickknacks—in the hopes that fixing the damage would somehow put an end to the problem. It never did.

By the time I was a teenager, I had started trying to intervene in the violence between my parents. I never actually saw my father hit my mother, but I knew he did from the bruises I sometimes saw. I remember coming out to the living room one night in the middle of one of those fights and seeing my father, drunk, with his gun in his mouth, saying he was going to kill himself. My brother came out, too, and the two of us talked him into putting the gun down. It's a wonder he survived those years.

Sometimes I think if my father hadn't been a police officer, he would have been a criminal. He used to tell a story about when he was a young cop, answering a false alarm at a tire store in the middle of the night. His more experienced partner opened the trunk of the police car, took out a spare tire, and flung it through the window of the store. Then they loaded all the new tires they could fit into the police car, drove to his house, dumped the tires on the lawn, and went back to the store for another load. They called all the other police officers on duty to come over and get in on the looting. Eventually, they called the owner and told him, "Your tire store was robbed."

In spite of my father's behavior, I respected him, even idolized him in some ways when I was young. As bad as your parents may be at their worst, they are the only parents you'll ever have. My dad was good-looking and charming when he was sober, and to me he seemed just like a TV detective, a larger-than-life figure hunting down bad guys and meting out justice. At the time, I didn't realize he was probably just another blue-collar guy, getting through the week to get to the weekend, getting through the years to get to retirement. Some people seem to need conflict, thrive on it, and create conflict everywhere they go. I've heard it said that the children of conflict seekers are raised to have the emotional control their parents lack and then some—that fighters raise peacemakers.

My parents bought a series of boats, always in deplorable condition. We would take them out into the Atlantic Ocean, well past the horizon. We'd go out in any kind of weather, sometimes straight into a blinding fog. We had no navigation equipment other than a compass and no working radio. We'd fish all day, and when we felt it was time to come back in, we would try to follow the charter fishing boats back into the inlet. When we lost them, because those boats were invariably faster than ours, we'd head west until we saw land, then head up or down the coast until we saw something we recognized. Often, our crappy engine would break down and we would drift until we could flag down another boat, one with a radio, to call the Coast Guard to tow us in. Sometimes we would even be taking on water, in danger of sinking. Each time, we'd get home, congratulate ourselves for surviving, and head right back out again as soon as we could. It never occurred to us that we should stop taking these risks, because we always survived by our wits, always seemed to learn something from it.

WHEN I WAS about eleven, my mother decided to become a cop. She'd done catering or babysitting occasionally to make extra money throughout my childhood, and then she'd become a secretary, which was unrewarding and didn't pay well. Now she wanted a career. The local police department had opened up the entrance exam to women, as many departments did in the 1970s. A lot of male police officers would have felt threatened by the thought of their wives trying to become officers as well. But not my father. To his credit, he encouraged her.

My mother studied for the civil service exam, which took time and effort. After she passed that, she had to take a physical fitness exam. She would have to meet all the same benchmarks as the men, and for a small woman, this was an enormous challenge. My father helped her set up an obstacle course in our backyard where she could practice every day. She ran around a set of cones carrying a toolbox filled with weights. She practiced dragging me a hundred feet across the backyard (in place of the dummy she would have to drag in the real test).

The toughest part was the wall she would have to scale, seven feet four inches. Knowing that, my father built a practice wall a bit higher than the real one. At first, she couldn't touch the top. It took her a long time before she was able to jump up and grab the top of the wall. Eventually she was able to pull herself up and get a leg over, and by honing this technique in practice sessions every day, she got to where she could scale that wall on the first try every time. The day of the test, she actually scaled the wall better than most of the men. She became one of very few women to pass the test, and that made a big impression on Mark and me: she had decided on a goal that seemed like it might not be possible, and she had achieved it through sheer force of determination and the support of people around her. I still hadn't found a goal for myself that would give me that same kind of drive, but I had at least seen what that would look like.

My memories of school are largely of being trapped in a classroom, bored out of my mind and always wondering what was going on outside. For my entire K–12 education, I pretty much ignored my teachers and daydreamed. I didn't know what I wanted to do, just that it would be exceptional, and I was pretty sure it had nothing to do with history, grammar, or algebra. I couldn't concentrate on any of it anyway. I was reading way behind grade level when I was seven, so my parents asked my maternal grandmother, who was a special education teacher, to evaluate me and try to help. After working with me for a few days, she gave up and declared me hopeless.

If I was a kid now, I imagine I would be diagnosed with ADHD. But back then, I was just a bad student. I learned to squeak by on whatever native intelligence I had, even though I never did any homework. My brother remembers a day in high school when our father sat us down and explained that he could get us into a welders' union when we graduated. He figured a trade would probably be our best option for a career because we were such poor students. Mark realized then that if he wanted to do something with his life more exciting or lucrative than welding, he had better improve his grades. So he got his ass in gear and did, starting that day. I have no recollection of this conversation, as I was probably looking out the window at a squirrel.

Meanwhile, the principal of our high school, Jerry Tarnoff, was begging me not to quit my trigonometry course, attempting to impress upon me that I had potential if I could just focus. I tried to explain to him how impossible it was for me to pay attention in that class, in any class. His words had no effect on me. I quit trigonometry. After that, whenever I saw him in the hallways, I would avoid his gaze. I was surprised by how much it bugged me to know I had let him down. Still, he never seemed to give up on me. Years later, he came to both of my space shuttle launches, and I think it meant a lot to him to see that his faith had paid off for at least one student.

The only thing that I could really connect with enough to succeed at was my work as an EMT. Mark worked with the local volunteer ambulance unit, too. Later, our dad twisted some arms (maybe literally) to get us hired at the paid ambulance service in nearby Orange, which was a rougher town than West Orange. We got the chance to see more kinds of medical emergencies and learned from them. I spent the summer after high school working as an EMT in Jersey City, which was like being thrown into the big leagues. I had found something that was meaningful to me and that I was good at. I decided to become a doctor, and I knew I could be a good one if I could just get through the ten years of training.

By screwing up my college application, I had wound up at the University of Maryland, Baltimore County. (I had meant to apply to the College Park campus.) In my freshman year, I started out with great hope that I could turn things around and be a good student, as I had every previous school year. This determination always lasted just a few days, until I realized once again that it was impossible for me to concentrate in class or to study on my own. Soon I was waking up each morning and struggling to think of a reason to go to class, knowing I wouldn't absorb any of the professor's lecture. Often, I didn't go. How was I going to graduate, let alone do well enough to be accepted by any medical school?

Everything changed that afternoon when I picked up *The Right Stuff.* I'd never read anything like it before. I'd heard the word "voice" used to describe literature, but this was something I could actually hear in my head. *Even out in the middle of the swamp,* Wolfe wrote, *in this rot-bog of pine trunks, scum slicks, dead dodder vines, and mosquito eggs, even out in this great overripe sump, the smell of "burned beyond recognition" obliterated everything else.* I felt the power of those words washing over me, even if some of the words I had to look up in the dictionary. *Perilous, neophyte, virulent.* I felt like I had found my calling. I wanted to be like the guys in this book, guys who could land a jet on an aircraft carrier at night and then walk away with a swagger. I wanted to be a naval aviator. I was still a directionless, undereducated eighteen-year-old with terrible grades who knew nothing about airplanes. But *The Right Stuff* had given me the outline of a life plan.

3

PAUL MCCARTNEY IS singing over the crackly communication system. So far we've heard Coldplay, Bruce Springsteen, Roberta Flack. I happen to like "Killing Me Softly," but I can't help but think it's inappropriate considering the circumstances. I'm crammed in the right-hand seat of the Soyuz, acutely aware of the 280 tons of explosive propellant under me. In an hour, we will tear into the sky. For now, soft rock is distracting us from the pain of sitting in the cramped capsule.

When we got off the bus at the launch site, it was fully dark, floodlights illuminating the launch vehicle so it could be seen from miles around. Though I've done it three times before, approaching the rocket I was about to climb into is still an unforgettable experience. I took in the size and power of this machine, the condensation from the hyper-cooled fuel billowing eerily in a giant cloud, enveloping our feet and legs. As always, the number of people around the launchpad surprised me, considering how dangerous it is to have a fully fueled rocket—basically a bomb—sitting there. At the Kennedy Space Center, the area was always cleared of nonessential personnel for three miles around, and even the closeout crew drove to a safe viewing site after strapping us into our seats. Today, dozens of people were milling around, some of them smoking, and a few of them will watch the launch from dangerously close. Once, I watched a Soyuz launch while serving as backup for one of the crew, I was standing outside the bunker, just a few hundred yards away. As the engines ignited, the manager of the launchpad said in Russian, "Open your airway and brace for shock."

In 1960, an explosion on the launchpad killed hundreds of people, an incident that would have caused a full investigation and an array of new regulations for NASA. The Soviets pretended it hadn't happened and sent Yuri Gagarin to space the following year. The Soviet Union acknowledged the disaster only after the information about it was declassified in 1989.

By tradition, there is one last ritual: Gennady, Misha, and I climb the first few stairs heading toward the elevator, then turn to say goodbye to the assembled crowd, waving to the people of Earth one last time.

Now we wait in the Soyuz, something we've all experienced before, so we know our roles and know what to expect. I anticipate the excruciating pain in my knees that nothing seems to alleviate. I try to distract myself with work: I check our communication systems and introduce oxygen into the capsule with a series of valves—one of my primary responsibilities as the flight engineer 2, a position that I like to describe as the copilot of the copilot of the spacecraft. Gennady and Misha murmur to each other in Russian, and certain words jump out: "ignition," "dinner," "oxygen," "whore" (the all-purpose Russian swear). The capsule heats up as we wait. The music we hear now is "Time to Say Goodbye" by Sarah Brightman, who was going to travel to the International Space Station later this year but has had to cancel her plans. A Russian pop song, "Aviator," follows.

The activation of the launch escape system wakes us up with a loud thunk. The escape system is a separate rocket connected to the top of the spacecraft, much like the one on the old Apollo/Saturn that was designed to pull the capsule free in case of an explosion on the pad or a failure during launch. (The Soyuz escape rocket was used once, saving two cosmonauts from a fireball, in 1983.) The fuel and oxidizer turbo pumps spin up to speed with a screaming whine—they will feed massive amounts of liquid oxygen and kerosene to the engines during ascent.

Russian mission control warns us it's one minute to launch. On an American spacecraft, we would already know because we'd see the countdown clock ticking backward toward zero. Unlike NASA, the

Russians don't feel the drama of the countdown is necessary. On the space shuttle, I never knew whether I was really going to space that day until I felt the solid rocket boosters light under me; there were always more scrubs than launches. On Soyuz, there is no question. The Russians haven't scrubbed a launch after the crew was strapped in since 1969.

"*My gotovy,*" Gennady responds into his headset. We are ready.

"*Zazhiganiye,*" mission control says. Ignition.

The rocket engines of the first stage roar to full capacity. We sit rumbling on the launchpad for a few seconds, vibrating with the engines' power—we need to burn off some of the propellant to become light enough to lift off. Then our seats push hard into our backs. Some astronauts use the term "kick in the pants" to describe this moment. The slam of acceleration—going from still to the speed of sound in a minute—is heart pounding and addictive, and there is no question that we are going straight up.

It's night, but we wouldn't be able to see anything out our windows even if it were broad daylight. The capsule is encased in a metal cylinder, called a fairing, which protects it from aerodynamic stress until we are out of the atmosphere. Inside, it's dark and loud and we are sweaty in our Sokol suits. My visor fogs up, and I have trouble reading my checklist.

The four strap-on boosters of four engines each fall away smoothly after two minutes, leaving the four remaining engines of the second stage to push us into space. As we accelerate to three times the Earth's gravity, the crushing force smashes me into my seat and makes it difficult to breathe.

Gennady reports to the control center that we are all feeling fine and reads off data from the monitors. My knees hurt, but the excitement of launch has masked the pain some. The second-stage rockets fire for three minutes, and as we are feeling their thrust, the fairing is jettisoned away from us in two pieces by explosive charges. We can see outside for the first time. I look out the window at my elbow, but I see only the same black we launched into.

Suddenly, we are thrown forward against our straps, then slammed back into our seats. The second stage has finished, and the third stage has taken over. After the violence of staging, we feel some roll oscillations, a mild sensation of rocking back and forth, which isn't alarming. Then the last engine cuts off with a bang and there is a jolt, like a minor car crash. Then nothing.

Our zero-g talisman, a stuffed snowman belonging to Gennady's youngest daughter, floats on a string. We are in weightlessness. This is the moment we call MECO, pronounced "mee-ko," which stands for "main engine cutoff." It's always a shock. The spacecraft is now in orbit around the Earth. After having been subjected to such strong and strange forces, the sudden quiet and stillness feel unnatural.

We smile at one another and reach up for a three-handed high-five, happy to have survived this far. We won't feel the weight of gravity again for a very long time.

Something seems out of the ordinary, and after a bit I realize what it is. "There's no debris," I point out to Gennady and Misha, and they agree it's strange. Usually MECO reveals what junk has been lurking in the spacecraft, held in their hiding places by gravity—random tiny nuts and bolts, staples, metal shavings, plastic flotsam, hairs, dust—what we call foreign object debris, and of course NASA has an acronym for it: FOD. There were people at the Kennedy Space Center whose entire job was to keep this stuff out of the space shuttles. Having spent time in the hangar where the Soyuz spacecraft are maintained and prepared for flight, and having observed that it's not very clean compared to the space shuttle's Orbiter Processing Facility, I'm impressed that the Russians have somehow maintained a high standard of FOD avoidance.

The Soyuz solar arrays unfurl themselves from the sides of the instrumentation module, and the antennas are deployed. We are now a fully functional spacecraft in orbit. It's a relief, but only briefly.

We open our helmets. The fan noise and pump noise blending together are so loud we have trouble hearing one another. I had remembered this about my previous mission to the ISS, of course, but still I can't believe it's so noisy. I can't believe I'll ever get used to it.

"I realized a few minutes ago, Misha," I say, "that our lives without noise have ceased to exist."

"Guys," Gennady says. "*Tselyi god!*" An entire year!

"*Ne napominai, Gena,*" answers Misha. Gena, don't remind me.

"*Vy geroi blya.*" You're freaking heroes.

"Yep," Misha agrees. "Totally screwed."

Now we are in the rendezvous stage. Joining two objects in two different orbits traveling at different speeds (in this case, the Soyuz and the ISS) is a long process. It's one we understand well and have been through many times, but still it's a delicate maneuver. We pick up a strange broadcast over Europe:

... scattered one thousand four hundred feet. Temperature one nine. Dew point one seven. Altimeter two niner niner five. ATIS information Oscar ...

It's some airport's terminal broadcast, a recording giving pilots information about weather and approaches. We shouldn't be receiving this, but the Soyuz comm system is horrible. Every time Russian mission control talks to us we can hear the characteristic *dit-dit-dit* of cell phone interference. I want to yell at them to turn off their cell phones, but in the name of international cooperation I don't.

A few hours into the flight my vision is still good, with no blurring—a positive sign. I do start to feel congested, though, which is a symptom I've experienced in space before. I feel my legs cramp, from being crammed into this seat for hours, and there is the never-ending knee pain. After MECO, we can unstrap ourselves, but there isn't really anywhere to go.

Gennady opens the hatch to the orbital module, the other habitable part of the Soyuz, where the crew can stay if it takes more than a few hours to get to the station, but this module doesn't have much more space. I disconnect my medical belt, a strap that goes around my chest to monitor my respiration and heartbeat during launch, and float up to the orbital module to use the toilet. It's nearly impossible to pee while still halfway in my pressure suit. I can't imagine how the women do it. After I get back into my seat, mission control yells at me to plug my medical belt back in. We get strapped back in a few hours before we will be docking. Gennady scrolls through the checklist on his tablet and starts inputting commands to the Soyuz systems. The process is

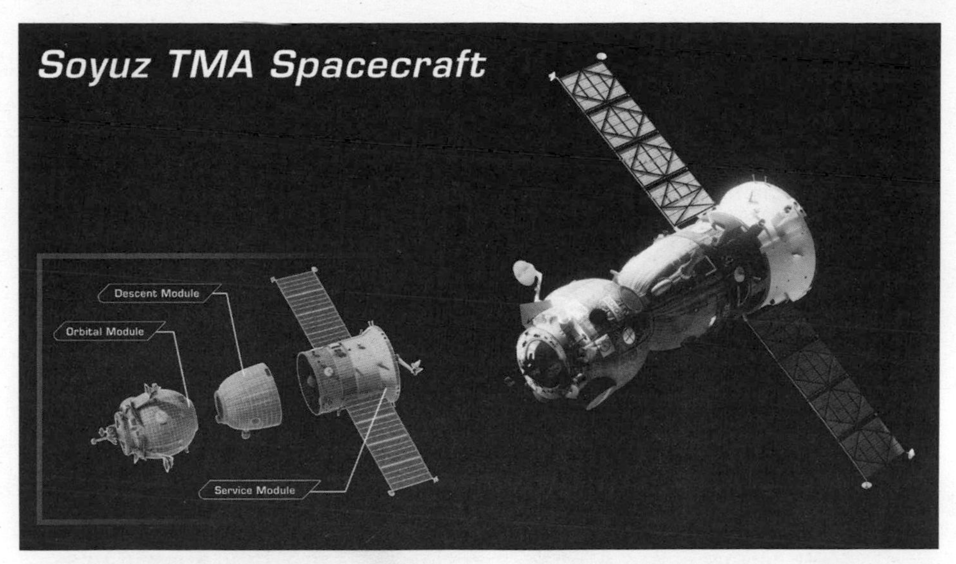

Soyuz rendering showing the three sections of the spacecraft: orbital module, descent module, service module

largely automated, but he needs to stay on top of it in case something goes wrong and he has to take over.

When it's time for the docking probe to activate, nothing happens. We wait. Gennady says something to Russian mission control in rapid-fire Russian. They respond, sounding annoyed, then garble into static. We are not sure if they heard us. We are still a long way from ISS.

"Fucking *blya,*" Gennady groans. Fucking bullshit.

Still no indication the docking probe has deployed. This could be a problem.

The process of docking two spacecraft together has remained pretty much unchanged from the Gemini days: one spacecraft sticks out a probe (in this case, us), inserts it into a receptacle called a drogue in the other spacecraft (the ISS), a connection is made, everyone cracks sex jokes, we leak-check the interface before opening the hatch and greeting our new crewmates. The process has been reliable for the past fifty years, but this time the probe doesn't appear to have worked.

The three of us give one another a look, an international I-can't-fucking-believe-this look. Soon, ISS will be looming in the window, its eight solar array wings glinting in the sun like the legs of a giant insect. But without the docking probe, we won't be able to connect to it and climb aboard. We'll have to return to Earth. Depending on when the next Soyuz will be ready, we might have to wait weeks or months. We could miss our chance altogether.

We contemplate the prospect of coming back to Earth, how ridiculous we'll feel climbing out of this capsule, saying hello again to people we've just said the biggest good-bye in the world to. Comm with the ground is intermittent, so they can't help us much in our efforts to figure out what's going on. I turn to see Misha's face. He is shaking his head in disappointment.

Once Gennady and Misha transition the computer software to a new mode, we see that the probe is in fact deployed. It was just a software "funny."

All three of us sigh with relief. This day hasn't been for nothing. We are still going to the space station.

I watch the fuzzy black-and-white image on our display as the docking port on ISS inches closer and closer. I wonder if it's true that the probe is actually okay. The last part of the rendezvous is exciting, much more dynamic than the space shuttle docking ever was. The shuttle had to be docked manually, so it was a slow ballet with little room for error. But the Soyuz normally docks with ISS automatically, and in the last minutes of the approach it whips itself around quickly to do an adjustment burn. Even though we'd known to expect this, it's still attention getting, and I watch out the window as the station comes flying into view, its brilliant metal sparkling in the sunlight as if it's on fire. The engines fire briefly, and we hear and feel the acceleration. Leftover fuel vents outside, glinting in the sun. With the burn complete, we snap back into position to move toward the docking port.

When we finally make contact with the station, we hear and feel the eerie sound of the probe hitting, then scratching its way into the drogue, a grinding metal-on-metal sound that ends with a satisfying clunk. Now both ISS and the Soyuz are commanded to free drift—they are no longer controlling their attitude and are rotating freely in space until a more solid connection can be made. The probe is retracted to draw the two vehicles closer together, then hooks are driven through the docking port to reinforce the connection. We've made it. We slap one another on the arms.

I join Gennady in the orbital module, where we struggle out of the Sokol suits we put on nearly ten hours earlier. We are tired and sweaty but excited to be attached to our new home. I take off the diaper I've been wearing since I left Earth and put it in a Russian wet trash bag for later disposal on the ISS. I get into the blue flight suit I call my Captain America suit because of the huge American flag emblazoned across the front. I hate these flight suits—the Russian who has been making them for years can't be made to understand that we stretch an inch or two in space, so within a few weeks I will no longer be able to wear the Captain America suit without having my balls crushed.

As eager as we are to greet our new crewmates, we need to make sure the seal between the Soyuz and ISS is good. The leak checks take nearly two hours. The space between the two docking compartments has to be filled with air, which we then test to determine whether its pressure is dropping. If it is, we don't have a good seal, and opening the hatch will cause ISS and Soyuz to lose their atmosphere. Occasionally, as we wait, we hear the crew on the other side banging on the hatch in a friendly greeting. We bang back.

The leak check finally complete, Gennady opens the hatch on our side. Anton Shkaplerov, the only cosmonaut on board the ISS, opens the Russian hatch on their side. I smell something strangely familiar and unmistakable, a strong burned metal smell, like the smell of sparklers on the Fourth of July. Objects that have been exposed to the vacuum of space have this unique smell on them, like the smell of welding—the smell of space.

There are three people up here already: the commander and the only other American, Terry Virts (forty-seven); Anton (forty-three); and an Italian astronaut representing the European Space Agency, Samantha Cristoforetti (thirty-seven). I know them all, some much better than others. Soon, we will all know one another much better.

I've known Terry since he was selected as an astronaut in 2000, though we haven't overlapped much in our work. Anton and Samantha I've only gotten to know well since we've been preparing for this mission over the last year. The last time I hung out with Anton was in Houston, before my last flight. We both got pretty drunk at my neighborhood bar, Boondoggles, and later ended up spending the night at a friend's house nearby since neither of us was in shape to drive.

Over the course of this year in space, Misha and I will see a total of thirteen other people come and go. In June, a Soyuz will leave with Terry, Samantha, and Anton, to be replaced by a new crew of three in July. In September, three more will join us, bringing our total to nine—an unusual number—for just ten days. Then, in December, three will leave, to be replaced a few days later. Misha and I hope that the change in crew members will help break up the mission and the monotony to make our year less challenging.

Unlike the early days of spaceflight, when piloting skill was what mattered, twenty-first-century astronauts are chosen for our ability to perform a lot of different jobs and to get along well with others, especially in stressful and cramped circumstances for long periods of time. Each of my crewmates is not only a close coworker in an array of different high-intensity jobs but also a roommate and a surrogate for all humanity.

Gennady floats through the hatch first and hugs Anton. These greetings are always jubilant—we know exactly who we're going to see when we open the hatch, but still it's somehow startling to launch off the Earth, travel to space, and find friends already living up here. The big hugs and big smiles you see if you watch the hatch opening live on NASA TV are completely sincere. As Gennady and Anton say their hellos, Misha and I are waiting our turns. We know that many people on the ground are watching, including our families. There is a live feed playing for everyone at Baikonur, as well as in mission control in Houston and online. The video signal is bounced off a satellite and then down to Earth, as with all of our communications. Suddenly I get an idea and turn to Misha.

"Let's go through together," I suggest. "As a show of solidarity."

"Good idea, my brother. We are in this together."

It's a bit awkward floating through the small hatch together, but the gesture gets a big smile from everyone on the other side. Once we're through, I shake hands with Anton.

Next I give Terry Virts a hug, then Samantha Cristoforetti. She is the first Italian woman to fly in space, and soon she'll be the record holder for the longest single spaceflight by a woman.

Our families in Baikonur are waiting to have a conference call with us, which we'll do from the Russian service module. I float down there and make a wrong turn. It's weird to be back here—floating through the station is so familiar, but it's also disorienting. It's only day one.

As the smell of space dissipates, I'm starting to detect the unique smell of the ISS, as familiar as the smell of my childhood home. The smell is mostly the off-gassing from the equipment and everything else, which on Earth we call the new car smell. Up here the smell is stronger because the plastic particles are weightless, as is the air, so they mingle in every breath. There is also the faint scent of garbage and a whiff of body odor. Even though we seal up the trash as well as we can, we only get rid of it every few months when a resupply craft reaches us and becomes a garbage truck after we empty it of cargo.

The sound of fans and the hum of electronics are both loud and inescapable. I feel like I have to raise my voice to be heard above the noise, though I know from experience I'll get used to it. This part of the Russian segment is especially loud. It's dark and a bit cold as well. I feel a shiver of realization: I'm going to be up here for nearly a year. What exactly have I gotten myself into? It occurs to me for a moment that this might be one of the stupider things I've ever done.

When we reach the service module, I notice right away that it's much brighter than when I was here last. Apparently the Russians have improved their lightbulbs. It's also much better organized than I remember, which I suspect is a result of Anton trying to impress Gennady with his organizational skills. Gennady is a stickler for keeping the Russian segment neat and tidy.

During the conference call, our families can see and hear us, but we can only hear them. There is a loud echo. The comm configuration up here is slightly off. I hear Charlotte telling me what the launch was like, then I talk briefly to my daughter Samantha and then to Amiko. It's great to hear their voices. But I'm conscious that my Russian colleagues are waiting to talk to their families, too.

Once we finish the call, I head down to the U.S. segment with Terry and Samantha Cristoforetti, where I'm going to spend the better part of the year to come. Though ISS is all one facility, for the most part the Russians live and work on their side and everyone else lives and works on the other side—"the U.S. segment." I notice it's much darker than I remember—burned-out lightbulbs haven't been replaced. This isn't Terry and Samantha's fault, but a reflection of the conservative way the control center has come to manage our consumables since I was last here. I decide to make it a project over the coming months to improve how we use our resources, since I'm going to be up here for so long, and good lighting will be critical to my well-being.

Terry and Samantha show me around, reminding me how things work up here now. They start with the most important piece of equipment to master: the toilet, also known as the Waste and Hygiene Compartment, or WHC. We also run through a quick safety brief that we will redo more thoroughly in a couple of days once I'm more settled. An emergency could strike at any time—fire, ammonia leak, depressurization—and I'll have to be ready to deal with whatever comes, even on day one.

We head back to the Russian segment for a traditional welcoming party—special dinners are held there every Friday night and on other special occasions, including holidays, birthdays, and good-bye dinners before each Soyuz leaves. Welcoming parties are one of those occasions, and Terry has warmed up my favorite, barbecued beef, which I stick to a tortilla using the surface tension of the barbecue sauce (we eat tortillas because of their long shelf life and lack of crumbs). We also have the traditional foods we share at Friday night dinners—lump crabmeat and black caviar. Everyone is in a festive mood. It's been a long, tough day for the three of us who just arrived. Technically, two days. Eventually we say our good nights, and Terry, Samantha, and I head back to the U.S. segment.

I find my crew quarters, or CQ, the one part of the space station that will belong just to me. It's about the size of an old-fashioned phone booth. Four CQs are arranged in Node 2: floor, ceiling, port side, starboard side. I'm on the port wall this time; last time I was on the ceiling. The CQ is clean and empty, and I know that over the course of the next year it will fill with clutter, like any other home. I zip myself into my sleeping bag, making a special point to appreciate that it's brand new. Though I will replace the liner a couple of times, the bag itself won't be cleaned or replaced over the next year. I turn off the light and close my eyes. Sleeping while floating isn't easy, especially when you're out of practice. Even though my eyes are closed, cosmic flashes occasionally light up my field of vision, the result of radiation striking my retinas, creating the illusion of light. This phenomenon was first noticed by astronauts during the Apollo era, and its cause still isn't thoroughly understood. I'll get used to this, too, but for now the flashes are an alarming reminder of the radiation zipping through my brain. After trying unsuccessfully to sleep for a while, I bite off a piece of a sleeping pill. As I drift off into a restless haze, it occurs to me that this is the first of 340 times I will have to fall asleep here.

4

FOR THE REST of that fall of 1982, I walked around the campus of the University of Maryland–Baltimore County with a new outlook on life. Before, I had always wondered where everyone got the motivation to get out of bed early to make it to class. They left parties while the music was still playing and while there were still beers unopened. Now I knew why: they each had some kind of goal. Now I'd found mine too, and it was a great feeling. I counted myself lucky to have picked up a book that showed me my life's goals so clearly, and I intended to fulfill them. Not only was I going to become a Navy pilot, I might even become an astronaut. These were the most challenging and exciting goals I had ever come across, and I was ready to get started. I had only one problem: the path to becoming a naval aviator is an extremely competitive one, and I was still a chronic underperformer with a terrible academic record. I would have to become a commissioned officer in the Navy, but that pipeline was clogged with accomplished young people who had excelled in high school and were then nominated to the U.S. Naval Academy by their congressman or senator. They had aced their SATs. Because I'd daydreamed and bullshitted my way through high school, I didn't have the basic knowledge even to begin the kinds of courses I'd have to take: calculus, physics, engineering. Beyond that, I knew that even if I started at a remedial level, I still probably wouldn't be able to keep up. However strong my motivation, I lacked the skills necessary to learn.

Everywhere I looked, I saw students who could listen to an hour-long lecture, asking intelligent questions and writing things down. They turned in homework assignments on time, correctly done. They took a textbook and lecture notes and did something they called "studying." They were then able to do well on exams. I had no idea how to do any of this. If you've never felt this way, it's hard to express how awful it is.

By now, my brother was a freshman at the U.S. Merchant Marine Academy in Kings Point, New York. Our maternal grandfather had been a Merchant Marine officer in World War II and later served as a fireboat captain for the New York City Fire Department. Mark was thinking of following in his footsteps in the Merchant Marine, but he wasn't dead set on it, so he liked the idea that the education he was getting at the academy was a good starting point for a range of careers. Given my new goals, this seemed like a good place to start for me too, because Kings Point offered a path to a commission in the Navy. Even if I couldn't get into a military academy, Kings Point would still give me the structure of a military environment, which I felt I needed. Best of all, I would already know someone there who could help me get my bearings as a transfer student. I arranged a meeting with an admissions counselor over the Christmas break.

When I arrived on campus that January wearing the most formal thing I owned—khaki pants and a polo shirt—I was greeted by the dean of admissions himself, in full military dress uniform. I'd never dealt with a uniformed officer before (aside from cops, of course). He invited me into his large office that seemed to be constructed almost entirely from wood—wood furniture, wood bookshelves, wood chairs, model ships, and other nautical memorabilia all over the walls. A tarnished brass ship's engine order telegraph stood alone in the far corner of the room. The dean looked me in the eye and asked me why I wanted to transfer there.

"Well, sir, I want to become a commissioned officer in the Navy. My goal is to fly fighter planes and land on aircraft carriers."

In my mind, this was such a clear and compelling goal. But the man's eyes glazed over as I spoke, and he kept looking at his watch, as if he was already thinking about his next appointment, or maybe about what he was going to have for lunch. He kept looking behind me, toward the window, rather than meeting my eyes. When I was done talking, he cleared his throat and closed the folder on his desk that contained my sad credentials.

"Look," he said, and sighed.

Not a good sign.

"Your high school grades are pretty terrible. Your SAT scores are below the average for our incoming freshmen. Your grades in your first semester of college are no better than high school. There's just nothing here to indicate that you would be successful in the very challenging program here."

"I intend to improve my grades now," I explained. "I know I can do it. And my SAT—I didn't even study for it. I think I could do much better if I tried taking it again."

"Well, your two scores would have to be averaged," he explained, "so you'd need a perfect score in order to bring your new score up to our average. And even that wouldn't be enough to balance out your grades."

This wasn't quite how I had expected this conversation to go.

I told him about growing up as the son of two cops, about my parents' deplorable boats, about working as an EMT. I told him about reading *The Right Stuff* and realizing what I wanted to do, finally finding a clear direction in life. I told him about the jets and the aircraft carrier and the risk and the possibility of achieving something important. I told him I thought it could all start at Kings Point. I asked him what I could do to change his answer.

He just shook his head. "I'm sorry, son," he said. "There's just no way with this record. You wouldn't succeed here." He stood, thanked me for coming in, and explained that it was only in deference to my brother, who was doing extremely well, that he had agreed to see me. He shook my hand and showed me the door.

Back outside in the bright sun, I blinked and looked around, stunned. I wasn't going to join my brother here, and I wasn't going to get started on the next step of my life. I was closer to tears than I'd been since I could remember.

I realize now that that dean of admissions must have spent a lot of his time listening to young people describe lofty goals that they had neither the talent nor drive to achieve, and to him I must have seemed no different from them. Maybe I wasn't. I can put myself in his position now, but at the time his indifference was devastating. Kings Point had seemed to be my only possibility: I could only assume I would meet rejection anywhere else I tried. Of course I couldn't get into Annapolis. I didn't have a plan B. Meanwhile, other people my age with the same goals were surging forward, and I would be years behind them. It seemed like it would take me a long time just to get to the starting gate—maybe so long the Navy wouldn't want me anymore. I knew there was an age limit for being commissioned into the Navy.

Everything else I'd done in my life up to this point, like working as an EMT, had been choices that had played to my strengths and hadn't particularly challenged my weaknesses. This new goal was going to expose every weakness I had.

In my second semester at UMBC, I signed up for more challenging classes and tried to apply myself for the first time in my life. I remember walking into class on the first day of precalculus—more or less the same material as the trigonometry class Mr. Tarnoff had begged me not to drop in high school—and thinking, *This is it. If I can't show what I'm capable of in this class, I'm not going to get the chance to do much more.* During my first semester I'd taken the easiest math class, algebra, in order to satisfy a requirement, and I'd barely passed it. Now I was in a class that would build on the knowledge I had already failed to absorb, and I had to do much, much better.

After the first class, I sat down to do the homework, feeling the pressure of everything I had now decided to do. I had to force myself to stay in my chair. I kept thinking of something else I needed to do in another room, a reason to walk down the hall of my dorm. I needed to sharpen my pencil. I needed a glass of water. I stayed in the chair anyway. I forced myself to read through the chapter, over and over. It still didn't make a lot of sense because a lot of the terms were ones I was supposed to have learned in high school and hadn't. I forced myself to work through the homework problems, and while I was pretty sure I had gotten the right answers to the easier ones, the harder questions were still pretty fuzzy. It was late at night by the time I was done, and I tried not to reflect on the fact that everyone else in my class had probably ripped through the homework in fifteen minutes. I tried to focus on the fact that I had set myself a goal—to read this chapter, to do the problems—and I had done that. I turned out the light feeling like I might finally be able to turn things around.

A few weeks later, I was doing my homework and realized that it was getting a *little* easier. It was still a lot like banging my head against a wall, but some of the material that I had struggled with the week before now seemed to make a little more sense. The whole process of learning became a little less painful, a little more reliable with each problem I slogged through. It was still a mighty struggle to stay in the chair, and I only got a B– in the class. Still, that B– was one of the major achievements of my life thus far. I had decided to learn something hard, and I had learned it.

Meanwhile, I applied to transfer to two schools: Rutgers University and the State University of New York Maritime College. Both were close by and both offered me the possibility of becoming a commissioned officer in the Navy.

SUNY Maritime is a small military-oriented school in the Bronx, established to train ships' officers in the maritime industry. It was the first maritime college in the country, built on Fort Schuyler (named for General Philip Schuyler, Alexander Hamilton's father-in-law). The fort was designed to defend Manhattan from naval attack in the aftermath of the War of 1812. I didn't know any of this when I applied there—it was just one of very few options that seemed open to me. When the school made me an offer of admission, I immediately accepted. I had just barely made the cutoff for the minimum GPA.

I didn't know it at the time, but I'd been rejected by Rutgers. When the letter came in the mail, my parents had opened it, then thrown it away without telling me. They couldn't bear to disappoint me. They kept this secret for years, until long after I'd been selected as an astronaut.

I made my way to Fort Schuyler in the summer of 1983. I knew I would start the year with a two-week indoctrination program, but I had only a hazy idea based on movies of what this would be like: getting my head shaved, upperclassmen shouting in my face, forced marches, having to clean things like shoes and belt buckles over and over. As it turned out, all of these ideas were entirely accurate.

THE SUNY MARITIME CAMPUS WAS surprisingly beautiful, on a spit of land between the Long Island Sound and the East River, under the Throgs Neck Bridge. Sprawling and well kept, the campus centered on the stately structure of the old fort, with newer buildings around the perimeter. On move-in day, I had only a footlocker full of clothes, a boom box, and cassette tapes of Journey, Bruce Springsteen, the Grateful Dead, and Supertramp. I found my room, a beige cube crammed with two single beds, two desks, and two dressers. My roommate was already there, unpacking. He introduced himself as Bob Kelman (the college did everything alphabetically, so Kelly and Kelman would room together and line up together for everything). Bob was—and still is—a friendly and outgoing person with a wry smile and a cutting sense of humor. We talked a bit as we unpacked, getting to know each other.

"So what are you going to do when you get out of here?" Bob asked.

"I'm going to be an astronaut," I said without a trace of a smile, looking him dead in the eye. I was trying to get used to taking the idea seriously myself. Bob narrowed his eyes and looked me up and down.

"Oh, yeah?" Bob asked.

"Yep," I responded, deadpan.

He nodded thoughtfully. "Well, I'm going to be an Indian chief."

He laughed pretty hard at his own joke. At the time I thought he was being sort of a jerk, but once we became good friends this story of his reaction always made us laugh, especially once I actually became an astronaut.

As we unpacked our things, Bob and I talked about the indoctrination period about to begin and what exactly that would be like. I made a joke about getting our heads shaved.

"What?" Bob froze, a pile of books in his hand. "They're not going to shave our heads. You're joking, right?"

I told him I was pretty sure it was true. "It's like a military indoctrination. Don't they always shave your head in the military?"

Bob thought about it for a moment, then dismissed it. "Nah," he said. "They would have told us. I mean, we would have had to sign something."

THE NEXT MORNING, Bob and I were awakened at five in the morning by upperclassmen beating on pots and pans and garbage can lids and screaming in our faces. We had five minutes to go from a dead sleep to dressed in our PT gear, beds made, standing in the hall at attention. That morning, and every morning to follow, started with an hour of running and calisthenics. It was already hot and sticky even before dawn, and once the sun came up the heat was brutal.

Those first days, we memorized quotations and phrases connected to the school's history. The first one we learned was the Sallyport Saying, which was inscribed over an archway in the old fort: "But men and officers must obey, no matter at what cost to their feelings, for obedience to orders, instant and unhesitating, is not only the life-blood of armies but the security of states; and the doctrine, that under any conditions whatever deliberate disobedience can be justified is treason to the commonwealth—Stonewall Jackson." (Basically, "Obey orders." If Jackson had been more succinct, my indoctrination would have been a lot easier.) I preferred a shorter, more compelling quotation: "The sea is selective, slow at recognition of effort and aptitude, but fast at sinking the unfit." I still remember those quotations to this day.

That first morning, we were marched to another building, where we were taken into a small room one by one. Because we did everything in alphabetical order, I got to watch Bob's reaction while I sat in the chair and had my head shaved. I didn't mind about my hair, but I can still remember the look on Bob's face, an expression of abject horror. I laughed so much that the guy with the clippers had to yell at me to hold still. A few minutes later, Bob's black curls were on the floor along with mine.

The military discipline came pretty easily for me. I think I had been craving that kind of structure, and it was almost a relief to be told what to do and how to do it. Many of my classmates questioned the logic and fairness of every aspect of our training, tried to cut corners, and whined and complained. But I had started to figure out that I needed a clear challenge in order to apply myself. Schoolwork was still hard for me, but following directions gave me stability. I embraced it.

At the end of the indoctrination period, we had a ceremony to mark the achievement. My parents came, dressed in their Sunday best, as did my paternal grandparents. As we marched by in formation, I saw all of them in the stands looking proudly at me. I was surprised by how much it meant to me to have them there, to have given them something to be proud of. But I was also aware of how far I still had to go.

When the school year started in earnest, I was taking six classes. I was nearly starting over as a freshman because the curriculum for the program here was so different from the random handful of arts and sciences classes I'd taken in Maryland. I was taking calculus, physics, electrical engineering, seamanship, and military history. The curriculum was challenging even for my classmates who had excelled in high school, and I felt good about the fact that I was keeping my head above water.

When Labor Day weekend approached, I got a call from one of my high school buddies inviting me to a party at their frat house at Rutgers. I said I'd be there.

I called my brother. "Let's go down to Rutgers and hang out with Pete Mathern at Sigma Pi," I said.

"I can't," Mark said right away. "I have a test coming up."

I spent a few minutes trying to talk him into it before he interrupted.

"Don't you have some sort of test coming up, too? You've been in classes for a few weeks now."

"Yeah," I admitted. "My first calculus exam is at the end of next week. But I'll study for it after I get back. I'll have Tuesday, Wednesday, Thursday . . ." In my mind, I was already on the side of the Cross Bronx Expressway with my duffel slung over my shoulder, thumb stuck out.

"Are you out of your goddamn mind?" Mark asked. "You're in *school.* You need to absolutely ace this exam, and everything else, if you want to get caught up. You need to spend this entire weekend at your desk, doing every problem in every chapter this exam is going to cover."

"Seriously?" I asked. "The entire long weekend?" This sounded insane to me.

"All weekend," he said. "*And* the whole coming week, too."

There was a weird silence while I took this in. I didn't appreciate being yelled at by my twin brother. It was tempting to tell myself he was just being a jerk and that I should ignore him. I came so close to deciding not to listen; the memory still unsettles me, like a memory of teetering on the edge of a cliff. As much as I wanted to go to the party, I knew somewhere in my mind that he was right and that he was offering me something important by being as blunt as he was. Mark had also started out as a distracted, indifferent high school student. But he had decided to pull himself together long before I did and had succeeded. I'd never asked him how he'd done it, but now he was trying to offer me the lesson of his experience. I reluctantly decided to listen.

I stayed in all weekend—hard as it was for me—and worked every problem in every chapter, just as he had suggested, until I could do them all. When I took the exam that Friday, I felt for the first time in my life as though I understood every question and thought I had answered them more or less correctly. It was a strange feeling. When we got our exams back the next week, there was a circled red 100 at the top of mine. I stared at it for the longest time, trying to reinforce to myself the sequence of events that had played out. I had earned a perfect score on a test for the first time in my life, and a math test to boot. This was how people got good grades. It was like a door had opened.

From that point on, I enjoyed the challenge of school. I knew how to work hard and enjoyed seeing it pay off. It almost became a game I played with myself: let's see how well I can do at this. In a strange way, it was easier for me to get an A+ than it would have been to get a B. Shooting for a lower grade was like aiming an arrow at a smaller target. "Just okay" is like threading a needle; "the best I can possibly do" is a much broader set of goalposts. I decided to try to know *everything.* Then I would always get an A.

That phone call with Mark was almost as pivotal a moment in my life as reading *The Right Stuff.* The book had given me a vision of who I wanted to be; my brother's advice showed me how to get there.

Soon after classes got started, I had gone into the ROTC office and said I wanted to join the unit. I learned that I could participate in their courses and training, but I couldn't apply for a scholarship until after I had at least one semester's worth of grades. So I trained with the other cadets, doing drills and weekend exercises and taking classes on leadership, weapons systems, and military etiquette. On top of all that, all Maritime students had to study to be licensed with the U.S. Coast Guard, which was a requirement to become a merchant mariner. (I did wind up getting my U.S. Coast Guard license and have kept it current to this day.) We learned celestial and terrestrial navigation, seamanship, meteorology for mariners, and nautical "rules of the road." After my first semester with a nearly 4.0 GPA, I was offered a Navy ROTC scholarship in exchange for at least five years of military service—longer if I went to flight school. I was pleased to be that much closer to my ultimate goal, and of course my parents were pleased that the rest of my tuition would be paid for.

At the end of the school year, we spent a few weeks preparing our training ship for our first cruise. The *Empire State V,* the former USNS *Barrett,* was a retired troop transport ship that we were learning to operate—we each had assigned tasks on the ship. When we finally started moving, I was standing watch on the bow, and the gray East River opened out before us as we crept away from the dock and headed into the fog of the Long Island Sound. I kept peering out intently into the thick soup as if the ship and every life on board depended on me: it had been drilled into me that the bow watch was not just the eyes of the ship but also the ears—I was listening for other ships and ready to call up to the bridge if I saw or heard anything that might pose a threat. As the engine room came up to speed, the distinct smell of the boiler oil tinged the air. I stood watch past City Island, past New Haven, and on to Montauk. Later, as we rounded the point to head east out into the North Atlantic Ocean, I took a deep breath of the ocean air. We were at sea. I felt like I was finally getting somewhere. I had the sense this was the stepping-off point for what would be many exciting adventures of discovery. I wouldn't be mistaken.

I STILL COULDN'T quite believe we were sailing to Europe. If you'd told me, a little more than a year earlier, that this was how I would spend my nineteenth summer, I wouldn't have believed you. The accommodations aboard the ship were dark, dingy, and poorly ventilated. When I headed up to the mess deck for meals, I often came across people throwing up into the large trash cans that lined the room. At night, people moaned in their bunks from nausea. I seemed to be immune to seasickness, and I hoped the vestibular fortitude would carry over into flight and eventually into weightlessness.

We worked on a three-day cycle: maintain the ship one day, stand watch one day, and attend classes one day. The best watch to draw was helm, as we would actually get to have a hand on the wheel steering the ship. Bow watch meant just looking out over the water trying to identify other ships. Stern watch meant looking out for someone falling overboard, which no one ever did. Class days we'd pile into small rooms filled with high school desks for academic instruction. Some of this was interesting—navigation, meteorology, and emergency procedures like firefighting or search and rescue. It wasn't my intention to become a ship's officer, but it seemed like a good backup career, so I paid attention and did as well as I could. At night we honed our celestial navigation skills, learning to fix the position of the ship using a sextant to measure the angle between the horizon and a particular star or planet. There was complicated math involved and it was tricky to learn, but it was necessary for sailing (and, I would learn later, for spaceflight too).

The first port we pulled in to was Majorca, Spain. (Beautiful beaches.) Then came Hamburg. (The only souvenir I have is a complete and permanent aversion to apple schnapps.) Next, we stopped at Southampton, England, and took a train into London. (I was shocked by how awful the food was for such a large cosmopolitan city.)

On the cruise back to home port, I felt I was getting the hang of the duties on board and the classroom material we were studying. We came back to Maritime stronger, more resilient, more competent than when we'd left. We'd learned to work together in difficult circumstances, responded to the unexpected, and survived. I'd understood that the point of the cruise had been to teach us seamanship, leadership, and teamwork, but it still surprised me how much I had learned. I stepped off the *Empire State V* a different person than when I'd stepped on.

AS SOON as I finished that first cruise, I got on a plane to Long Beach, California, to do ROTC training on a Navy ship cruising to Hawaii. I was with midshipmen from other colleges, including the Naval Academy, who were doing all of this for the first time. Though I was only a few months ahead of them in experience, it seemed like much more.

This was my first real exposure to the Navy. Freshman ROTC and naval academy midshipmen were expected to do the work of enlisted sailors, so when I became an officer leading enlisted men, I would know what their responsibilities were like. I lived in crowded berthing again, with about twenty guys in bunks stacked three high. It was good practice for living in small spaces in the future. Just as on the *Empire State V*, we did a lot of manual labor on the ship, which some guys resented. But I wasn't bothered by it and was happy to be on a ship, making progress toward a career in the Navy.

. . .

MY SECOND CRUISE on the *Empire State V*, the summer between my sophomore and junior years, offered me better work assignments and more authority. Our first night in port in Alicante, Spain, my classmates and I threw a party in our room. Drinking was not allowed on the ship, but as long as we didn't cause any problems we hoped not to get in any trouble. Within a few hours we were pretty lit. I finished off a bottle of vodka, the last of the alcohol we had on board, and I thought I should mark the occasion by throwing the bottle against the bulkhead to smash it. But instead of breaking, the bottle bounced off the wall and struck one of my classmates on the back of the head. She was nearly knocked out, and we probably should have sought medical attention for her, but instead we thought it was hysterical, including her.

Intent on continuing our party, we came up with a great plan: we decided to get a Jacob's ladder (a hanging ladder made of heavy rope and wood planks) and throw it over the back of the ship. Then we could climb down, swim to the dock, and sneak off to a nearby bar. We dispatched a couple of people to the forward part of the ship to find the ladder and haul it back. When they reached the rest of us, waiting for them at the stern of the ship, they were dragging the ladder, which weighed nearly a hundred pounds. As we were putting it into place, I got into an argument with a classmate about which of us would go down the ladder first. We were yelling and screaming, neither of us backing down, and nearly came to blows. I finally convinced him of my superior qualifications for the job and triumphantly climbed over the railing to test how securely the ladder was fastened. In fact, it wasn't tied down at all. I fell, along with one hundred pounds of rope and wood, thirty feet into the dark water below. I remember hitting the hard, cold water as if hitting a sheet of pavement and reflecting with surprise that I had remained conscious. I quickly sank, pulled down by the heavy ladder, which I was now tangled in. It took a huge effort to swim back up to the surface. I was able to struggle over to the engineering side port, used for loading supplies while the ship was docked, and some of the engineering cadets were already there waiting to pull me back in. I was completely limp from the shock of hitting the water so hard, as well as from the vodka, but eventually a classmate pulled me in through the access door. I made it back to the aft part of the ship undetected, and our superiors never learned of our adventures. I surely would have been expelled if they had, and that would have cost me the one chance I had managed to create for myself.

5

April 3, 2015

Dreamed I was working in a Soviet-era car refurbishment plant with Soviet soldiers, wearing their olive-green full-length wool coats and their Russian hats. The plant took old crappy Soviet cars and cleaned them up, maybe for resale and maybe for some other nefarious purpose. I wasn't sure. I was responsible for cleaning the engines with a big steamer. Each time I sprayed, engine oil splattered all over the room, and I worried that I was somehow doing it wrong. I wondered how the room would be cleaned.

MY GOAL FOR most of my adult life has been to pilot aircraft and spacecraft. So it sometimes strikes me as odd that the International Space Station doesn't need to be piloted at all. When I try to explain this to people who don't know much about the station, I tell them it's more like a ship traveling the world's oceans than like an airplane. Something like the USS *La Jolla*, a submarine I served on for a few days as a midshipman while still in college, which was self-contained and self-powering. We don't fly the space station—it's controlled by software, and even if human intervention is needed, it is controlled by laptops on board or on the ground. We live in the space station, the way you live in a building. We work in it, the way scientists work in a laboratory, and we also work *on* it, the way mechanics work on a boat, if the boat were adrift in international waters and the Coast Guard had no way to reach it.

I sometimes see the station described as an object: "The International Space Station is the most expensive object ever created." "The ISS is the only object whose components were manufactured by different countries and assembled in space." That much is true. But when you live inside the station for months, it doesn't feel like an object. It feels like a *place*, a very specific place with its own personality and its own unique characteristics. It has an inside and an outside and rooms upon rooms, each of which serve different purposes, its own equipment and hardware, and its own feeling and smell, distinct from the others. Each module has its own story and its own quirks.

I've been on station for a week now. I'm getting better at knowing where I am when I first wake up. If I have a headache, I know it's because I've drifted too far from the vent blowing clean air at my face. I'm often still disoriented about how my body is positioned—I'll wake up convinced that I'm upside down, because in the dark and without gravity, my inner ear just takes a random guess as to how my body is positioned in the small space. When I turn on a light, I have a sort of visual illusion that the room is rotating rapidly as it reorients itself around me, though I know it's actually my brain readjusting in response to new sensory input.

The light in my crew quarters takes a minute to warm up to full brightness. The space is just barely big enough for me and my sleeping bag, two laptops, some clothes, toiletries, photos of Amiko and my daughters, a few paperback books. Without getting out of my sleeping bag, I wake up one of the two computers attached to the wall and write down what I remember of my dreams. After my last flight, people were interested in my descriptions of the vivid and surreal dreams I had in space, but I forgot most of them, so I'm making a point of keeping a more consistent dream journal this time.

Then I look at my schedule for today. I click through new emails, stretch and yawn, then fish around in my toiletries bag, attached to the wall down by my left knee, for my toothpaste and toothbrush. I brush,

My crew quarters on board the ISS, a place to sleep but also my personal private space. My home in our home away from home for a year

still in my sleeping bag, then swallow the toothpaste and chase it with a sip of water out of a bag with a straw. There isn't really a good way to spit in space. I spend a few minutes looking over the daily summary sent up by the Mission Control Center in Houston, an electronic document that shows the state of the space station and its systems, asks us questions they came up with overnight, and includes important notes for the plan we're going to execute that day. There's also a cartoon at the end, often making fun of either us or themselves. Today's daily summary shows it's going to be a challenging day, and these are the days I look forward to.

Mission control schedules our days into increments as short as five minutes using a program called OSTPV (Onboard Short Term Plan Viewer), which rules our lives. Throughout the day, a dotted red line moves relentlessly across the OSTPV window on my laptop, pushing through the block of time mission control has estimated for each task. NASA people are optimists by nature, and unfortunately this optimism can extend to the estimate of how long it will take me to perform a certain task, such as repairing a piece of hardware or conducting an experiment. If I take longer than scheduled to complete a task, the extra time has to come out of something else on the schedule—a meal, my exercise time, the brief time I get to myself at the end of the day (which OSTPV labels "pre-sleep"), or—worst of all—sleep. Most of us wind up having complicated relationships with the line on the OSTPV screen. Sometimes when I'm working on something challenging, the line seems to speed up malevolently, and I could swear something is wrong with it. Other times, it seems to settle down and match the passage of time as I perceive it. Of course, if I could somehow zoom out my view of the schedule wide enough to take in the entire year, the line would be creeping forward so slowly it wouldn't appear to be moving at all. Today's schedule seems well thought out, but there are a few ways in which things could go wrong. For Terry, Samantha, and me, much of today is to be taken up with one long task labeled DRAGON CAPTURE.

From the outside, the International Space Station looks like a number of giant empty soda cans attached to one another end to end. The length of the station is made up of five modules connected the long way—three American and two Russian. More modules, including ones from Europe and Japan as well as the United States, are connected as offshoots to port and starboard, and the Russians have three that are attached up and down (we call these directions zenith and nadir). Between my first mission to the space station and this one, it has grown by seven modules, a significant proportion of its volume. This growth is not haphazard but reflects an assembly sequence that had been planned since the beginning of the space station project in the 1990s. Whenever visiting vehicles are berthed here—resupply spacecraft like the Russian Progress, American Cygnus, Japanese HTV, or SpaceX Dragon—for a time there is a new "room," usually on the Earth-facing side of the station; to get into one of them, I have to turn "down" rather than turning left or right. Those rooms get roomier as we get the cargo unpacked, then get smaller again as we fill them with trash. Not that we need the space—especially on the U.S. segment, the station feels quite roomy, and in fact we can lose one another in here easily. But the appearance of extra rooms—and then their disappearance, after we set them loose—is a bit strange. It used to be that uncrewed cargo vehicles were built as one-use spacecraft, and after we detached them from the station they burned up in the atmosphere. The relatively new SpaceX Dragon has the capability to return to Earth intact, which gives us more flexibility.

I won't get to spend time outside the station until my first of two planned spacewalks, which won't be for almost seven months. This is one of the things that some people find difficult to imagine about living on the space station—the fact that I can't step outside when I feel like it. Putting on a spacesuit and leaving the station for a spacewalk is an hours-long process that requires the full attention of at least three people on station and dozens more on the ground. Spacewalks are the most dangerous thing we do on orbit.

Even if the station is on fire, even if it's filling up with poisonous gas, even if a meteoroid has crashed through a module and our air is rushing out, the only way to escape the station is in a Soyuz capsule, which also needs preparation and planning to depart safely. We practice dealing with emergency scenarios regularly, and in many of these drills we race to prepare the Soyuz as quickly as we can. No one has ever had to use the Soyuz as a lifeboat, and no one hopes to.

The space station is an international effort and a shared facility, but in practice I spend almost all of my time on the cluster of modules—which, together with American and Japanese visiting vehicles—we call the U.S. operational segment. My cosmonaut colleagues spend the majority of their time on the Russian segment, consisting of the Russian modules as well as the visiting Russian Progress and Soyuz spacecraft.

The module where I spend a lot of my day is the U.S. module formally named Destiny, but which we mostly just call "the lab." It's a state-of-the-art scientific laboratory with walls, floors, and ceiling packed with equipment. Because there is no gravity, every surface is usable storage space. There are science experiments, computers, cables, cameras, tools, office supplies, freezers—crap all over the place. The lab looks cluttered—people with OCD would probably have trouble living and working here—but the things I use most I can put my hands on in seconds. There are also a large number of things I would *not* be able to put my hands on if asked—without gravity, items wander off regularly, and the ground will often email us WANTED posters regarding lost objects, like the ones the FBI puts in post offices. Occasionally one of us will dislodge a tool or part that has been missing for years. Eight years is the record, so far, for a missing object reappearing.

Most of the spaces where I spend my time have no windows and no natural light but rather bright fluorescent lights and clinical white walls. Devoid of any earthly color, the modules seem cold and utilitarian, like a prison of sorts. Because the sun rises and sets every ninety minutes, we can't use it to keep track of time. So without my watch keeping me on Greenwich Mean Time and a schedule tightly structuring my days, I'd be completely lost.

It's hard to explain to people who haven't lived here how much we start to miss nature. In the future there will be a word for the specific kind of nostalgia we feel for living things. We all like to listen to recordings of nature—rainforests, birdcalls, wind in the trees. (Misha even has a recording of mosquitoes, which I think goes a bit too far.) As sterile and lifeless as everything is up here, we do have windows that give us a fantastic view of Earth. It's hard to describe the experience of looking down at the planet. I feel as though I know the Earth in an intimate way most people don't—the coastline, terrain, mountains, and rivers. Some parts of the world, especially in Asia, are so blanketed by air pollution that they appear sick, in need of treatment or at least a chance to heal. The line of our atmosphere on the horizon looks as thin as a contact lens over an eye, and its fragility seems to demand our protection. One of my favorite views of the Earth is of the Bahamas—a large archipelago with a stunning contrast from light to dark colors. The vibrant deep blue of the ocean mixes with a much brighter turquoise, swirled with something almost like gold, where the sun bounces off the sandy shallows and reefs. Whenever new crewmates come up to the station for the first time, I make a point of taking them to the Cupola

Looking down at Earth from the Cupola module on board ISS

(a module made entirely of windows looking down on Earth) to see the Bahamas. That sight always reminds me to stop and appreciate the view of the Earth I've been given the privilege of seeing.

Sometimes when I'm looking out the window it occurs to me that everything that matters to me, every person who has ever lived and died (besides the six of us), is down there. Other times, of course, I'm aware that the people on the space station with me are the whole of humanity for me now. If I'm going to talk to someone in the flesh, look someone in the eye, ask someone for help, share a meal with someone, it will be one of them.

SINCE BEFORE the space shuttle was retired, NASA has been contracting with private companies to develop spacecraft capable of supplying the station with cargo and, at some point in the future, new crews. The most successful private company so far has been Space Exploration Technologies, better known as SpaceX, which produces the Dragon spacecraft. Yesterday, a Dragon launched successfully from the Kennedy Space Center. Since then, Dragon has been in orbit a safe ten kilometers from us. This morning, our aim is to capture it with the space station's robot arm and attach it to the berthing port on station. The process of grappling a visiting vehicle is a bit like playing an arcade claw machine, except that it involves real equipment worth millions of dollars flying at impossible speeds. Not only could an error cause us to lose or damage the Dragon and the valuable supplies on board, but a slip of the hand could crash the visiting vehicle into the station. A Progress cargo spacecraft once struck the old Russian space station *Mir,* and its crew was lucky not to have been killed by the rapid loss of atmosphere.

These uncrewed rockets are the only means by which we can get adequate supplies from Earth. The Soyuz spacecraft has the capability to send three human beings to space, but there is almost no room left over for anything else. SpaceX has had a lot of success so far with their Dragon spacecraft and Falcon rocket, and in 2012 they became the first private company to reach the International Space Station. They hope to be able to fly astronauts on the Dragon in the next few years. If they can pull that off, they will be the first private company to carry human beings to orbit, and that launch will be the first time astronauts leave Earth from the United States since the space shuttle was retired in 2011.

Right now, Dragon is carrying 4,300 pounds of supplies we need. There is food, water, and oxygen; spare parts and supplies for the systems that keep us alive; health-care supplies like needles and vacuum tubes for drawing our blood, sample containers, medications; clothing and towels and washcloths, all of which we throw away after using them for as long as we can. Dragon will also be bringing new science experiments for us to carry out, as well as new samples to keep the existing ones going. Notable among the science experiments is a small population of twenty live mice for a study we will be conducting on how weightlessness affects bone, muscle, and vision. Each resupply spacecraft also carries small care packages from our families, which we always look forward to, and precious supplies of fresh food that we enjoy for just a few days, until it runs out or goes bad. Fruits and vegetables seem to rot faster here than on Earth. I'm not sure why, and seeing the process makes me worry that the same thing is happening to my own cells.

We are especially looking forward to this Dragon's arrival because another resupply rocket exploded just after launch back in October. That one was a Cygnus flown by another private contractor, Orbital ATK. The station is always supplied far beyond the needs of the current crew, so there was no immediate danger of running out of food or oxygen when those supplies were lost. Still, this was the first time a rocket to resupply ISS had failed in years, and it destroyed millions of dollars' worth of equipment. And the loss of vital supplies like food and oxygen made everyone think harder about what would happen if a string of failures were to occur. A few days after the explosion, an experimental space plane being developed by Virgin Galactic crashed in the Mojave Desert, killing the copilot, Michael Alsbury. These failures were unrelated, of course, but the timing made it feel as though a string of bad luck might be catching up with us after years of success.

I get dressed while I skim over the procedures for the Dragon capture again. We all trained for this thoroughly before launch, capturing many imaginary Dragons using a simulator, so I'm just refreshing my recollection. Getting dressed is a bit of a hassle when you can't sit or stand, but I've gotten used to it. The most challenging thing is putting on my socks without gravity to help me bend over. It's not a challenge to figure out what to wear, since I wear the same thing every day: a pair of khaki pants with lots of pockets and strips of Velcro across the thighs, crucial when I can't put anything down. I have decided to experiment with how long I can make my clothes last, the idea of going to Mars in the back of my mind. Can a pair of underwear be worn four days instead of two? Can a pair of socks last a month? Can a pair of pants last six months? I aim to find out. I put on my favorite black T-shirt and a sweatshirt that, because it's flying with me for the third time, has to be one of the most traveled pieces of clothing in the history of clothing.

Dressed and ready for breakfast, I open the door to my CQ. As I push against the back wall to float myself out, I accidentally kick loose a paperback book: *Endurance: Shackleton's Incredible Voyage* by Alfred Lansing. I brought this book with me on my previous flight as well, and sometimes I flip through it after a long day and reflect on what these explorers went through almost exactly a hundred years before. They were stranded on ice floes for months at a time, forced to kill their dogs for food, and nearly froze to death in the biting cold. They hiked across mountains that had been considered impassable by explorers who were better equipped and not half starved. Remarkably, not a single member of the expedition was lost.

When I try to put myself in their place, I think the uncertainty must have been the worst thing. The doubt about their survival would be worse than the hunger and the cold. When I read about their experiences, I think about how much harder they had it than I do. Sometimes I'll pick up the book specifically for that reason. If I'm inclined to feel sorry for myself because I miss my family or because I had a frustrating day or because the isolation is getting to me, reading a few pages about the Shackleton expedition reminds me that even if I have it hard up here in some ways, I'm certainly not going through what they did.

Out in Node 1, the module that serves largely as our kitchen and living room, I open a food container attached to the wall and fish out a pouch of dehydrated coffee with cream and sugar. I float over to the hot water dispenser in the ceiling of the lab, which works by inserting a needle into a nozzle on the bag. When the bag is full, I replace the needle with a drinking straw equipped with a valve to pinch it closed. It was oddly unsatisfying at first to drink coffee from a plastic bag sipped through a straw, but now I'm not bothered by it. I flip through the breakfast options, looking for a packet of the granola I like. Unfortunately, everyone else seems to like it too. I choose some dehydrated eggs instead and reconstitute them with the same hot water dispenser, then warm up some irradiated sausage links in the food warmer box, which resembles a metal briefcase. I cut the bag open, then clean the scissors by licking them, since we have no sink (we each have our own scissors). I spoon the eggs out of the bag onto a tortilla—conveniently, surface tension holds them in place—add the sausage and some hot sauce, roll

Terry Virts and me taking a break from the workday in Node 1, our living room and dining room onboard the ISS

it up, and eat the burrito while catching up with the morning's news on CNN. All the while I'm holding myself in place with my right big toe tucked ever so slightly under a handrail on the floor. Handrails are placed on the walls, floors, and ceilings of every module and at the hatches where modules connect, allowing us to propel ourselves through the modules or to stay in place rather than drifting away.

There are a lot of things about living in weightlessness that are fun, but eating is not one of them. I miss being able to sit in a chair while eating a meal, relaxing and pausing to connect with other people. Eating on the space station, at my workplace, three times a day, while constantly floating and steadying myself, is hardly the same. My egg burrito will float if I let go of it, as will my spoon, egg crumbs, a squeeze bottle of mustard that came up on the last resupply rocket, and a tiny perfect sphere of coffee. The "table" we use for eating has Velcro strips and duct tape to help us keep things in place, but it's still a challenge to manage all these potentially floating components. I bite the coffee sphere out of the air and swallow it before it can drift into a piece of equipment, or onto a crewmate or my pants (as they need to last six months). The biggest concern is food getting stuck on the hatch seal between modules, one of which is right by the table where we eat. We need to be able to close and seal that hatch quickly in an emergency.

As I'm eating, Terry floats in and wishes me a good morning while looking for coffee. Terry's astronaut class of 2000 has gotten a raw deal in terms of flight opportunities, since the *Columbia* disaster grounded the fleet at the same time they completed their initial training and became eligible to fly. Terry didn't get his chance on the shuttle for ten years. He served as the pilot for STS-130, the mission that delivered the last two modules to the International Space Station—Node 3 and the Cupola. Terry should have then had the chance to command a shuttle mission of his own, but the program ended soon after. He had to wait another four and a half years before flying again, on this mission.

Like me, Terry was a test pilot before joining NASA—in his case, with the Air Force. He has thick dirty-blond hair, a pleasant demeanor, and his default expression is a smile. His call sign is "Flanders," after the lovably square character Ned Flanders on *The Simpsons.* Terry has the positive attributes of Ned Flanders—optimism, enthusiasm, friendliness—and none of the negative ones. He is one of a small handful of vocally religious astronauts, and while some of my colleagues are bothered by this, I've never had a problem with Terry on this issue or any other. I've found him to be consistently competent, and as a leader, he is a consensus builder rather than an authoritarian. Since I've been up here and have been commander, he has always been respectful of my previous experience, always open to suggestions about how to do things better without getting defensive or competitive. He loves baseball, so there's always a game on some laptop, especially when the Astros or the Orioles are playing. I've gotten used to the rhythm of the nine-inning games marking time for a few hours of our workday.

Terry eats a maple muffin top while I'm finishing my egg burrito. Next, I eat a pouch of rehydrated oatmeal with raisins. The food portions are small, to discourage waste, so we often wind up eating a few different things for one meal. We are going to have a long morning, and I don't know when we'll be able to break for lunch.

My crewmates and I converge in the U.S. lab for the daily planning conference with mission control in Houston, people at other NASA sites, and their counterparts in Russia, Japan, and Europe. I find I'm adapting to being up here quicker than last time, both physically—in terms of living in weightlessness—and in terms of following the routines, using the equipment, doing the work. I have a different outlook this time knowing that I'm going to be here so long. I'm running a marathon rather than a sprint. As I pace myself for a year's stay, I have to constantly remind myself that for certain things, better can be the enemy of good enough.

The conference generally starts at 7:30 a.m. our time. I say good morning to Samantha, who is already there; Gennady, Misha, and Anton will take part in the conference from the Russian segment. Once we are all assembled, Terry grabs the microphone from its position Velcroed to the wall.

"Houston, station on Space to Ground One, we are ready for the DPC."

Mission control answers with a bright "Good morning, station!" even though it's 2:30 a.m. in Houston. We go over the day's plans for a few minutes, mostly about the details of the Dragon capture. We'd been given a general timeline, but now we nail down exactly what time we need to get started with the procedure, the status of Dragon, whether it's behaving as expected, and when it will be in certain positions relative to the station. When we are done with Houston, they hand us over to the Marshall Space Flight Center in Huntsville, Alabama. Then Huntsville hands us over to Munich so we can coordinate with the European Space Agency. Then we talk to "J-COM" in the Japanese mission control in Tsukuba, Japan. Then it's time to talk to Russia: Terry turns it over to the cosmonauts by saying "*Dobroye utro, Tsup va Moskvy, Anton pozhaluysta.*" ("Good morning, Control Center Moscow. Anton, please.") Then Anton takes over the mic because he's in charge in the Russian segment, and he leads the planning meeting with the Russians. Their meeting style is very different from ours—the ground asks the cosmonauts how they are feeling, which seems like a waste of time because they never say anything other than "*khorosho*"—good. At times I've dared them to say "not great," "just okay," or even "I feel like shit," but they refuse to take me up on it even when I offer them money.

The cosmonauts report on the atmospheric pressure of the station, information their flight controllers can see plainly on their own consoles. Next, they have to read back a list of deorbit parameters that, again, the ground already has—they sent them up to us. I find this waste of time maddening, but maybe it's an excuse to talk to the crew and gauge their moods and frustration levels.

The Russian space agency has a much different system for compensating their cosmonauts than we do: Their base salaries are much lower, but they get paid bonuses for each day they fly in space. (I get only five dollars per diem, but my base salary is much better.) However, their bonuses are decreased whenever they make "mistakes," those mistakes defined rather arbitrarily. I suspect that complaining, even making very legitimate complaints, can be defined as a mistake, costing them money and, potentially, the chance to fly in space again. As a result, everything is always "*khorosho.*"

All of this coordinating with sites all over the world might sound time-consuming, and it can be, but no one would ever suggest changing it. With so many space agencies cooperating, it's important that everyone knows what everyone else is doing. Plans can change quickly, and a misunderstanding could be costly, or deadly. We do this whole circuit of control centers both morning and evening, five days a week. I've chosen not to think about how many times I'll do these before I come back to Earth.

DRAGON IS in its orbit ten kilometers away from us, matching our speed of 17,500 miles per hour. We can see its light blinking at us on the external cameras. Soon, SpaceX mission control in Hawthorne, California, will move it to within two kilometers. Then authority transfers to mission control in Houston. There are stopping points along the way, at 350 meters, then 250 meters, then 30 meters, and finally the capture point at 10 meters. At each stopping point, teams on the ground will check Dragon's systems and evaluate its position before calling "go" or "no go" to move on to the next stage. Inside 250 meters, we will get involved by making sure the vehicle stays within a safe corridor, that it is behaving as expected, and that we are ready to abort if required. Once Dragon is close enough, Samantha will capture it with one of the station's robot arms. This is a glacially slow and deliberate process—one of the many things that's very different between movies and real life. In the films *Interstellar* and *2001: A Space Odyssey,* a visiting spacecraft zips up to a space station and locks onto it, a hatch pops open, and people pass through, all over the course of a few minutes. In reality, we operate with the knowledge that one spacecraft is always a potentially fatal threat to another—a bigger threat the closer it gets—and so we move very slowly and deliberately.

Samantha is going to operate the robot arm from the robotics workstation in the Cupola today (the robot arm's official name is Canadarm2 because it was made by the Canadian Space Agency). Terry will be her backup, and I will be helping out with the approach and rendezvous procedures. Terry and I squeeze in with her, watching the data screen over her shoulder that shows the speed and position of Dragon.

Samantha Cristoforetti is one of the few women to have served as a fighter pilot in the Italian Air Force, and she is unfailingly competent. She is also friendly and quick to laugh, and among her many other qualifications to fly in space, she has a rare talent for language. She has a native-level fluency in English and Russian, the two official languages of the ISS—she sometimes acts as an interpreter between cosmonauts and astronauts if we have to talk about something nuanced or complicated. She also speaks French, German, and her native Italian, and she's also working on learning Chinese. For some people who hope to fly in space, language can be a challenge. We all have to be able to speak at least one second language (I've been studying Russian for years, and my cosmonaut crewmates speak my language much better than I speak theirs), but the European and Japanese astronauts have the added burden of learning *two* languages if they don't already speak English or Russian.

When I first met Samantha, I would have described her as a hip young European; she has a worldly, sophisticated air about her. I found out later that she had participated in foreign exchange programs when she was young, attending high school in Minnesota for a year, spent a lot of time in Germany, and went to Alabama one summer when she was a teenager to attend Space Camp, a simulated astronaut training program. She has a seriously geeky side—she often tweets about science fiction like *Doctor Who* and *The Hitchhiker's Guide to the Galaxy*, and a lot of people were surprised and moved when she tweeted a picture of herself wearing a *Star Trek* uniform and giving a Vulcan signal out to the cosmos when Leonard Nimoy died. I am impressed with how well Samantha deals with the European Space Agency's control center in Munich; at times they seem indifferent and inattentive to what we are doing on board, which can be frustrating. She brings good humor to the most boring or annoying situations.

Before Samantha left Earth, she took Terry to her hair salon in Houston so her stylist could teach him to replicate her sleek asymmetrical haircut in space. Haircuts are one of the many tasks ISS crew members have to perform for one another (in addition to giving simple medical tests, drawing blood, doing ultrasounds, and even performing basic dentistry). Terry and Samantha posted pictures of the hairstyling lesson on Twitter, and their followers seemed tickled by the idea that Terry, the upcoming commander of the International Space Station, was being trained as a temporary cosmetologist. Halfway through their mission together, the big day came: Samantha felt her hair was getting too long and asked Terry to get out the equipment. Because we can't leave bits of hair floating in the air for others to inhale, our haircut equipment includes a vacuum cleaner. Terry tried his very best, but he still screwed it up—the layers that Samantha's stylist had made seem so easy to replicate under Earth's gravity were now floating all over the place. Samantha has spent the rest of her mission with her thick, dark hair sticking out from her head in a perfect brush that reminds me of a Russian fur hat.

The capcom speaking to us on the ground today is David Saint-Jacques, a Canadian astronaut. The term "capcom" is left over from the early days of Mercury when the astronauts went to space in capsules—one person in mission control was designated the "capsule communicator," the sole person in voice contact with the astronaut in space. "Capsule communicator" was shortened to "capcom," and the name stuck. Today David is talking us through the capture process, announcing Dragon's position as it moves, controlled by the ground through each of its preplanned stops.

"Station, Houston, on Space to Ground Two. Dragon is inside the two-hundred-meter keep-out sphere."

The keep-out sphere is an imaginary radius boundary around the station, meant to protect us from accidental collisions. "The crew now has the authority to issue an abort." This means that we can stop the process ourselves if we lose contact with Houston or if Dragon is outside the corridor. As Dragon approaches station, the early morning light catches the jagged edges of the Himalayas below. The Earth seems to zip by at an impossible speed.

"Houston on Two for rendezvous," Terry says. "Houston, capture conditions are confirmed. We're ready for Dragon capture. We're ready for step four."

"Copy that, stand by for capture, and just for your situational awareness, we expect that to take us about five or six minutes." Teams on the ground will give us one final go/no go for capture.

When the Dragon is within ten meters, we inhibit the station's thrusters to prevent any unintended jolts. Samantha takes control of the robotic arm, using her left hand to control the arm's translation (in, out, up, down, left, right) and her right hand to control its rotation (pitch, roll, and yaw).

"Station on Two for rendezvous," we hear from mission control. "You are go for capture sequence."

"Station copies," Samantha answers.

Samantha reaches out with the robot arm, watching a monitor that offers a view from a camera on the "hand," or end effector, as well as two other video monitors showing data describing Dragon's position and speed. She can also look out the big Cupola windows to see what she's doing. She moves the arm out away from the station—very slowly and deliberately. Closing the space between the two spacecraft inch by inch, Samantha never wavers or goes off course. On the center screen, the grapple fixture on Dragon grows larger and larger. She makes very precise adjustments to keep the spacecraft and the robot arm perfectly lined up.

The arm creeps out slowly, slowly. It's almost touching the Dragon.

Samantha pulls the trigger. "Capture," she says.

Perfect.

"We have nominal capture confirmed at 5:55 a.m. Central Time, while the station and Dragon fly over the northern Pacific Ocean, just to the east of Japan."

Samantha's round face has been a study of concentration, her bright brown eyes seeming almost not to blink. The moment capture is confirmed, her face relaxes into a huge smile and she high-fives Terry and me.

Terry speaks: "Houston, capture is complete. Samantha did a perfect job grappling Dragon."

"Copy and concur. Great job, guys. Congratulations."

Samantha takes the mic. "I just wanted to say thank you to the folks at SpaceX and you guys in Houston. It's been just amazing, watching the launch and knowing that it was headed our way and sure enough came knocking on our door. It was steady as a rock and we're very happy to have it here. It's exciting to have a new SpaceX dock. There's lots of science, and even coffee is in there. That's pretty exciting. So again, thanks a lot and great job to everybody."

"Thank you, Sam, and thank you, Terry, there's a bunch of very grateful people on the ground to see this go as smoothly as it has today. Nice job."

Now control is passing to the robotics officer in Houston (we call him "Robo"), who will maneuver Dragon into a position to be attached to the berthing port on the Earth-facing side of Node 2. Robo is controlling the robotic arm by typing in angles for the joints—they are analyzed by software to ensure the trajectory is safe before they are implemented. Once Dragon is lined up correctly, I will get involved again, monitoring when Dragon comes close enough to the station for "soft" berthing—four nine-inch latches reach out and grab Dragon and pull it into final contact with the ISS—followed by "hard" berthing, sixteen bolts driven through the connection between the space station and the Dragon to securely mate the two spacecraft.

The process of pressurizing the space between the Dragon and the station (the "vestibule") takes several hours and is important to do correctly. The danger Dragon poses to the station is not over: a mistake in this procedure could cause depressurization. So Samantha and I work through the steps one by one, making sure to do it right. First we check the integrity of the seal between the station and the Dragon by introducing air into the opening between them, a bit at a time. As when we arrived in the Soyuz, if the air pressure inside the vestibule were to decrease, even a tiny bit, that would indicate that the seal has been compromised and that opening the hatch would mean venting our breathable air out into the cosmos.

After a number of iterations of this process—introduce air, wait, measure pressure; repeat—we declare the seal safe, but we will wait until tomorrow to open the hatch. That step requires its own sequence of exacting steps. I've seen crews push themselves to get through the entire process because they were so eager to get into their care packages and fresh food. The process takes hours, though, and especially after the morning we spent with capture, it doesn't seem like a good idea to push things—there's too much risk of making a mistake. It will take us the next five weeks to unload all the cargo.

WHEN I FLOAT into my CQ for a moment to check my email, it's the first time I've had the chance to pause and think today. The carbon dioxide level is high today, nearly four millimeters of mercury. I can check it on the laptops and see exactly what the concentration of CO_2 is in our air, but I don't need to—I can feel it. I can sense the levels with a high degree of accuracy based only on the symptoms I've come to know so well: headaches, congestion, burning eyes, irritability. Perhaps the most dangerous symptom is impairment to cognitive function—we have to be able to perform tasks that require a high degree of concentration and attention to detail at a moment's notice, and in an emergency, which can happen anytime, we need to be able to do those tasks right the first time. Losing just a fraction of our ability to focus, make calculations, or solve problems could cost us our lives. And we are still learning about the long-term effects of breathing so much CO_2. It may cause cardiovascular problems and other issues in the future that we don't yet understand.

My tumultuous relationship with carbon dioxide has been going on as long as I've been flying in space. On my first shuttle mission, which was seven days, I was responsible for changing the lithium hydroxide canisters on board that scrubbed the CO_2 from the air. I remember each time I changed the canisters, once in the morning and once at night, I would become aware soon after how fresh the air was—only then did I realize that we had been breathing bad air. Part of our training in advance of flying on the shuttle was meant to let us experience and recognize the symptoms of high CO_2; we each went into a booth in the flight medicine clinic to put on a breathing mask that gave us slowly increasing levels of carbon dioxide.

On my second flight into space, again on the space shuttle, I became more aware of how CO_2 was affecting me and talked with my crewmates about their symptoms. I could pinpoint those moments when the CO_2 was the highest just from the way I felt. I decided to make an effort to start a more serious conversation about its effects. A new space station program manager had just been appointed, and soon after I was back on Earth I helped arrange to bring him on a visit to a Navy submarine under way in the Florida Straits. I thought the submarine environment would be a useful analogy for the space station in a number of ways, and I especially wanted my colleagues to get an up-close look at how the Navy deals with CO_2. What we learned on that trip was illuminating: the Navy has their submarines turn on their air scrubbers when the CO_2 concentration rises above two millimeters of mercury, even though the scrubbers are noisy and risk giving away the submarine's location. By comparison, the international agreement on ISS says the

CO_2 is acceptable up to six millimeters of mercury! The submarine's chief engineering officer explained to us that the symptoms of high CO_2 posed a threat to their work, so keeping that level low was a priority. I felt that NASA should be thinking of it the same way.

When I prepared for my first flight on the ISS, I got acquainted with a new carbon dioxide removal system. The lithium hydroxide cartridges were foolproof and reliable, but that system depended on cartridges that were to be thrown away after use—not very practical, since hundreds of cartridges would be required to get through a single six-month mission. So instead we now have a device called the carbon dioxide removal assembly, or CDRA, pronounced "seedra," and it has become the bane of my existence. There are two of them—one in the U.S. lab and one in Node 3. Each weighs about five hundred pounds and looks something like a car engine. Covered in greenish brown insulation, the Seedra is a collection of electronic boxes, sensors, heaters, valves, fans, and absorbent beds. The absorbent beds use a zeolite crystal to separate the CO_2 from the air, after which the lab

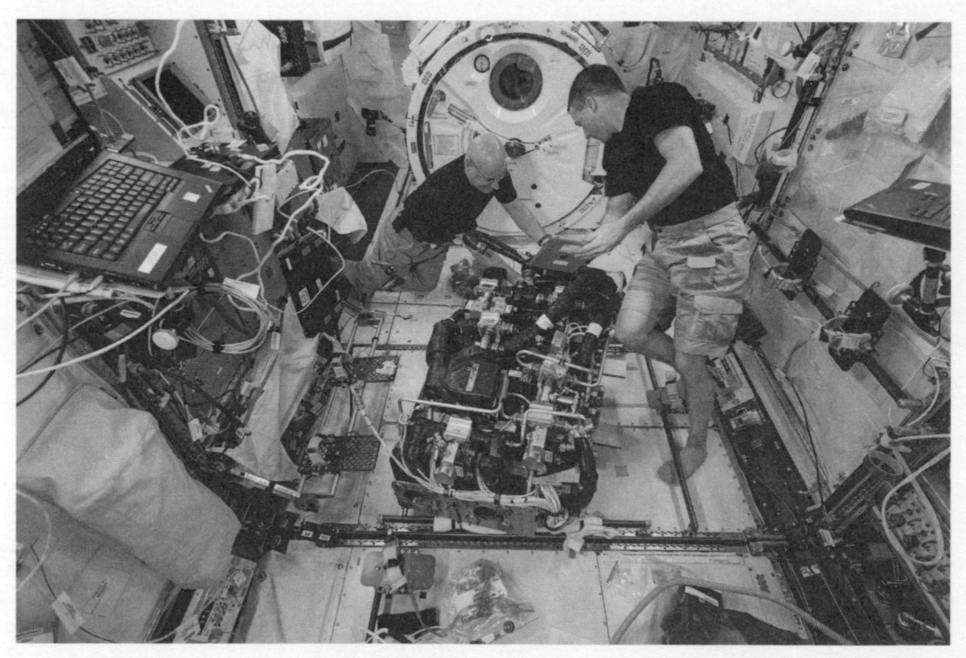

Terry Virts and me working on the Seedra in the Japanese module of the ISS

Seedra dumps the CO_2 out into space through a vacuum valve, while the Node 3 Seedra combines oxygen drawn from the CO_2 with leftover hydrogen from our oxygen-generating system in a device called Sabatier. The result is water—which we drink—and methane, which is also vented overboard.

The Seedra is a finicky beast that requires a lot of care and feeding to keep it operating. It wasn't until I was a month or so into my first mission aboard the space station that I started to correlate the symptoms I was feeling to specific levels of CO_2. At two millimeters of mercury I feel okay, but at around three I get headaches and start to feel congested. At four, my eyes burn and I can feel the cognitive effects. If I'm trying to do something complex, I actually start to feel stupid, which is a troubling way to feel on a space station. Above four millimeters of mercury, the symptoms become unacceptable. The levels can creep up for different reasons. Sometimes the Seedra has to be shut off because the space station's orientation isn't allowing us to collect enough solar energy to power it. For example, when a Progress resupply spacecraft docks, the solar arrays need to be turned edge on so their surface area doesn't get blasted by its thrusters. At other times there's no clear reason for Seedra going on strike. Sometimes, it's just broken.

Much of the management Seedra demands can be done from the ground, which is true for a lot of the hardware up here. Mission control can send a signal to the equipment, using the same satellites we use for email and phone calls. But at times, more serious hands-on maintenance by astronauts is required. The repair process isn't simple. Seedra has to be powered down and allowed to cool. Then all the electrical connectors, water-cooling lines, and vacuum lines at the bottom of the rack the Seedra sits in have to be removed. All of the bolts holding it in place must be removed so it can slide out. On my previous mission, when I gave Seedra a good tug, it didn't budge. It felt as if it were welded in place. I had to call to the ground for help, and they had no clue. Many meetings were called over the next few days at the Johnson Space Center while specialists tried to work through the problem.

In that instance, I went over all the bolts again and found one just hanging by a single thread. Problem solved. I pulled the beast out and eventually had to remove all the insulation to expose more electrical connectors, more water lines, and Hydro Flow connectors, which are notoriously tricky to mate. Working on a complex piece of hardware in space is infinitely harder than it would be on Earth, where I could put down tools and parts and they'd stay put. And there are so many complex pieces of hardware up here—NASA estimates that we spend a quarter of our time on maintenance and repairs. The hardest part of repairing the Seedra is replacing all the insulation, sort of like doing a huge 3-D puzzle with all the pieces floating. When we started it up again, it worked. Kelly 1, Seedra 0. I had no idea what it still had in store for me.

On this mission, the two Seedras have been giving us new issues to deal with. The one we use most, in Node 3, has been shutting down when its air selector valves, which are moving parts, get gummed up with zeolite and stick in the wrong position. The Seedra in the lab has an intermittent electrical short that we can't quite pin down. Sometimes, over the course of a day, the CO_2 level will slowly start to rise, especially when someone is exercising. As the day goes on, I'll feel congested, with burning eyes and a mild headache. I've been using Sudafed and Afrin to fight the symptoms, but these are temporary fixes, and I will quickly develop tolerances. A few days ago, I asked Terry and Samantha how they have been feeling, and they both said they had noticed that when the CO_2 was high they didn't feel especially sharp cognitively. I'm frustrated that we can't seem to get any urgency on this issue from the ground.

Part of my annoyance has to do with the fact that even though we have two Seedras on board, the ground allows us to run only one of them, keeping the other in reserve as a backup. We use the one in Node 3 because it works relatively consistently; only if it goes out, or if we have more than six people on board (as will happen in September), will we be authorized to use both. Our CO_2 level could come down to a much more tolerable level with the flip of a switch in Houston, and yet we can't convince them to do this. I can't help but to sometimes suspect the second Seedra is kept shut off to avoid the hassle of maintaining it from the ground. It's hard to work up sympathy for flight controllers who make this decision while breathing relatively clean Earth air. That level seems unconscionably high to me. The Russian managers claim that the CO_2 should be kept high deliberately because it helps to protect the crew from harmful effects of radiation. If there is any scientific basis for this claim, I have yet to see it. And because (I suspect) the cosmonauts are docked pay for complaining, they don't complain.

If we are going to get to Mars, we are going to need a much better way to deal with CO_2. Using our current finicky system, a Mars crew would be in significant danger.

THE LAST PLANNING CONFERENCE of the day will be held at 7:30 p.m., and dinnertime is shortly after that. As it's a Friday, we are looking forward to sharing a group dinner in the Russian segment, as always. Misha is usually the first one ready to start the weekend, and he floats over to the U.S. side in the afternoon to make a plan.

"What time should we start dinner, my brother?" he asks, his blue eyes wide and eager.

"How about eight?" I ask.

"Let's make it seven forty-five," he responds.

I agree.

After finishing up the DPC that evening and checking on an experiment, I give Amiko a quick call. "I'm heading over to Boondoggles," I tell her, jokingly referring to the Russian segment as our neighborhood bar in Houston. She understands what I mean. I start gathering things to bring to Friday dinner in a big ziplock bag. I pack my own spoon and my own scissors for opening food bags. I pack foods to share, stuff from my bonus food container and from what I brought up with me: canned trout, some irradiated Mexican meat, and a processed cheese similar to Cheez Whiz that Gennady loves. The Russians always share some tarry black caviar, for which I've developed a real taste, as well as some canned lobster meat. Samantha always brings good snacks, too—the Europeans have the best food.

With my goodie bag under my arm, I float into Node 1, then pass through the pressurized mating adapter (PMA-1), sort of a short, dark entryway between the U.S. and Russian segments. This entryway is not beautiful or spacious; it's about six feet long and canted up at a steep angle. It's quite narrow by design, and it's made even narrower by the cargo we store there in white fabric bags. I pass through the Russian module called the FGB (*funktsionalno-gruzovoy blok,* functional cargo block), then into the service module. There I find Gennady and Samantha watching a movie on a laptop while Anton floats horizontally to them, finishing up an experiment on the wall. On the laptop, a young woman's face flickers across the screen, a look of apprehension wrinkling her brow, while a man's voice speaks sternly in Russian.

"Hey, what are you guys watching?" I ask.

"It's *Fifty Shades of Grey,*" Samantha answers, "dubbed into Russian." In English, Gennady welcomes me and thanks me for the food I've brought, then in Russian tries to convince Samantha that *Fifty Shades* is a great literary work.

"That's ridiculous," Samantha says, without taking her eyes off the screen. She and Gennady argue, half-jokingly, about the literary status of *Fifty Shades* in rapid-fire Russian as Misha emerges from the bathroom. Terry shows up with his own goodie bag and greets everyone.

Anton welcomes us. He flew MiGs in the Russian Air Force before being selected as a cosmonaut, making him one of the people I might have found myself face-to-face with in combat had geopolitics in the early 1990s played out differently. He is solid and dependable, both physically and technically. He has a goofy sense of humor and is a close talker, even for a Russian. He has a halting way of speaking English with pauses in unusual places in his sentences, but I'm sure my Russian sounds far worse. I once asked Anton what he would have done if his MiG-21 and my F-14 had been flying straight at each other on some fateful day—how would he have maneuvered his airplane to get an advantage on me? When I was training and flying as a Navy fighter pilot, these questions about MiGs and their capabilities consumed my fellow pilots and me. All we knew then was guesswork based on military intelligence. As it turns out, the same guesswork was happening on the Soviet side. From Anton and the other cosmonauts I've gotten the impression that they didn't have much knowledge about our airplanes, and the training I got in dogfighting, flying against a very capable pilot in an F-16 pretending to be a MiG, was likely overkill by a wide margin. The Russian pilots are no less talented, they just had much less flight time than we did in our planes (I had more than 1,500 hours of experience in the F-14, while Anton has probably 400 hours in his MiG), presumably because their budgets were limited.

Anton and Misha acted as though Gennady were in charge as soon as he was on board, even though Anton is officially the Russian segment lead. Gennady has been, as always, awesome—things simply seem to go better when he is around, and everyone looks up to him as a natural leader. He doesn't do anything to try to grab power, but there is something about him that makes people want to listen to him.

Misha has been great to fly with so far too. He has a true concern for other people, and when he asks me regularly how I am doing, he really wants to know. He cares about what's going on in his friends' lives, how they are feeling, and what he can do to help. What's most important to him is friendship and camaraderie, and he brings esprit de corps to everything he does.

I'm often asked how well we get along with the Russians, and people never quite seem to believe me when I say there are no issues. People from our countries encounter cultural misunderstandings every day. To Russians, Americans can at first come across as naïve and weak. To Americans, Russians can seem stony and aloof, but I've learned this is just one layer. (I often think of a phrase I once read describing the Russian temperament as "the brotherhood of the downtrodden," the idea that Russians are bound by their shared history of war and disaster. I thought I read it in *The Master and Margarita* by Mikhail Bulgakov, but I've never been able to find it in any translation; maybe I read it in Russian and this was my own translation.) We make an effort to learn about and respect one another's cultures, and we have agreed to carry out this huge and challenging project together, so we work to understand and see the best in one another. The crewmates I fly with are crucial to nearly every aspect of my mission. Working with the right person can make the toughest day go well, and working with the wrong person can make the simplest task excruciatingly difficult. Depending on who is up here with me, my year in space could be needlessly perilous, fraught with conflict, or saturated with the everyday annoyance of a person you can't quite click with and also can't get away from. So far, I've been very lucky.

Once we're all gathered around the table, Gennady clears his throat and makes a solemn face that lets us know he's about to make a toast. The Russians are very formal about their toasts, and the first one of the evening is the most important. It's always to honor the people present and our reason for being together.

"*Rebyata*," he starts—"guys"—"can you believe we are here in space? The six of us are the only people here representing planet Earth right now, and I'm honored to be here with you. This is awesome. Let us drink to us and to our friendship."

"To us," the rest of us chime in, and the evening has officially begun.

It's challenging for six people to eat together in such a small space, but we look forward to this chance to have a meal as a crew. We use Velcro and duct tape to secure our dinners, but there is always some stray item—a drink bag, a spoon, a cookie—floating away from its owner and needing to be retrieved. It becomes part of the dining experience to reach out and grab someone's drink as it floats by your head. We listen to music while we eat, usually my playlist on the iPod I brought with me—U2, Coldplay, Bruce Springsteen. The Russians especially like Depeche Mode. Sometimes I'll sneak in some Pink Floyd or Grateful Dead. The Russians don't seem to mind my sixties rock, but they aren't very interested in hip-hop even though I've tried multiple times to introduce them to the works of Jay Z and Eminem.

We talk about how our work has gone during the week. The Russians ask about how the Dragon capture went, and we ask them about how the next Progress resupply schedule is looking. We talk about our families and catch up on current events in our respective countries. If there is significant news involving both the United States and Russia, for instance our two countries' involvement in Syria, we'll touch on it lightly, but no one wants to go into any detail. Sometimes the Russians will get caught up in an American news story. For instance, when two inmates escaped from a prison in upstate New York, Gennady and Misha were fascinated with them, asking me repeatedly whether they had been captured. I would find them lingering to watch updates on CNN on our projection screen whenever they had a reason to pass through Node 1.

As the evening goes on, the Russians make their second toast, which is often about something more specific, like current events. This toast is to Dragon and the supplies it brought us. The third toast is traditionally to our wives or significant others and our families. We all stop and think of our loved ones for a moment when Anton makes this toast.

We get to talking about what it's like to return to Earth in the Soyuz. Most of us have experienced this at least once before—Gennady has done it the most, four times—but for Terry and Samantha, their return in May will be their first time. It's a wild ride, and the four of us who have done it share our experiences. Gennady tells a story about one of his previous Soyuz flights, when the capsule hit the ground and then rolled around quite a bit, leaving the cosmonauts with their heads below their feet. One of Gennady's crewmates had tried to smuggle out some souvenirs in his pressure suit, and the extra cargo, along with the strange position they had landed in, left the unnamed cosmonaut with all his body weight on his groin. He was in so much distress that Gennady unstrapped himself, nearly breaking his neck when he landed on his head, in order to help reposition his crewmate and alleviate his pain. Terry and Samantha don't look too inspired by the story.

Friday night dinners always include dessert. Russian space dessert is almost always just a can of stewed apples. We have much more variety on the U.S. segment, though our desserts aren't gourmet level. The cherry blueberry cobbler is one of my favorites, and the chocolate pudding is always a big hit with the Russians, so I've brought some to share. It drives me nuts that our food specialists insist on giving us the same number of chocolate, vanilla, and butterscotch puddings, when the laws of physics dictate chocolate will disappear much faster. No one gets a vanilla craving in space (or on Earth).

We say our good nights and float back to the U.S. segment, remembering to bring our spoons and leftovers with us. Back in my CQ, I look through the plan for tomorrow, Saturday. As often happens up here, work will continue into the weekend, and I will do my required exercise sessions as well. I take off my pants and secure them under a bungee cord, don't bother changing my shirt, and brush my teeth. I put on my headset and call Amiko to talk for a few minutes before going to sleep. It's still early in the evening for her. I tell her about the Dragon capture, about *Fifty Shades of Grey*, about how the carbon dioxide is bugging me again, about Gennady's Soyuz story. She tells me about her workday, a lot of which she spent recording an episode of NASA's *Space to Ground* web series. Not long ago, she told me that her older son, Corbin, had advised her to take a break from thinking about the space station now and then. "Your work is space, and your home life is space," he told her. "You never get away from it." He was right. She is still helping her eighteen-year-old son, Tristan, deal with the consequences of his car having caught fire. She has also been helping my daughter Samantha and running errands for my father. I'm lucky to have Amiko taking care of things for me on the ground, and sometimes it bugs me that I can never do much to help her. This year in space is a test of endurance for Amiko as well, and it's important for me to remember that.

It's strange waking up here on weekends, even more so than waking up other days, because on weekends it becomes clearer that I'm sleeping at my workplace. I wake up Saturday and I'm still at work; wake up on Sunday, still at work. Months later, I'll still be here. On the weekends we are usually given time to do personal things—videoconference with family, catch up with personal email, read, get a little break from the relentless red line of OSTPV, and get the rest we need to start another week of long days of exacting work.

But there is a certain amount of mission creep into our time on weekends. The couple of hours of exercise on at least one day of the weekend are mandatory, since the damage to our bodies caused by weightlessness does not observe weekends or holidays. And there is station maintenance that can't be left until Monday, or that we won't have time to do when Monday comes around. The weekend is also when we clean, and cleaning is a bit more involved in zero gravity. On Earth, dust and lint and hair and fingernail clippings and bits of food *fall*, so dusting and vacuuming get rid of pretty much everything. On the space station, a piece of dirt can wind up on the wall, the ceiling, or attached to an expensive piece of equipment. A lot of crap winds up on the filters of the ventilation system, and when too much of it starts to build up, our air circulation is affected. Because the walls get dirty and wet, mold is a concern. And because mold spores don't fall to the floor but linger in our breathable air, they can pose a serious health risk. As a result, we are expected to clean most everything on station we regularly touch every weekend, with a vacuum and antiseptic wipes. We also take samples from the walls to grow in petri dishes and send back to Earth for analysis. So far they haven't found anything toxic, but it's both disgusting and fascinating to see what we are cultivating on the walls.

Then there's the Saturday morning science. When I was up here four years ago, there was an option to participate in additional science work on Saturday mornings, an idea introduced by an astronaut colleague who wanted to volunteer some of his free time to work on experiments that would otherwise go neglected. Since then, astronauts with a special interest in science could participate in Saturday morning science, and those who had other interests, or who—like me—felt they needed time to recover from the stress of the week in order to be ready for emergency situations, were under no pressure to do so. Now Saturday morning science no longer seems optional. In addition to all that, we need to start unpacking Dragon's cargo. Some of the cargo on Dragon is time sensitive (live mice and fresh vegetables, most notably). Once everyone is up and caffeinated, Terry and Samantha meet me in Node 2. We arm ourselves with checklists and cameras—a still camera to document each step of our work for later analysis by NASA and SpaceX, and a video camera so mission control can see exactly what we're doing in real time. When we're ready, we call down to the ground so they can follow along with us.

When Samantha opens the ISS hatch that leads to the Dragon and slides it out of the way, an unmistakable smell hits me—slightly burned, slightly metallic—the smell of space. Samantha smiles at me when she recognizes it. She has smelled it before, when her earlier crewmates went through a process similar to this one to open the hatch on her Soyuz, and again when two of her crewmates did a spacewalk.

We remove and stow a canvas covering that protects the hatch. Then Samantha and I work together to remove the four assemblies that power the latches and bolts that mated the two spacecraft together. It's a long, involved procedure to remove and properly cap all the connectors. The biggest risk here is damaging one of the connectors or losing a cap, troublingly easy to do when everything is floating around. We connect cables for power and data between the two spacecraft.

We tell the ground that we have successfully completed these steps.

"Station, Houston on Space to Ground Two, you have a go for step six, ingressing Dragon," capcom tells us.

"Copy that."

We put on goggles and dust masks before opening Dragon's hatch to protect us from dust and debris that might be floating inside. Samantha opens the hatch and slides it aside, then turns on the light inside Dragon. The first task is to make sure the air gets mixed between the two spacecraft—there is some danger that Dragon could be harboring a pocket of CO_2 or some other gas, and without gravity to keep the air constantly mixing, we have to install vent lines that will keep air circulating here as it does in the rest of the station. We take samples of the air inside Dragon to send back for analysis on Earth, and the Russians take their own sample (because NASA has sometimes questioned the Russian space agency's atmospheric standards, they insist on testing our air as well). We visually inspect the area around both hatches to make sure nothing has been damaged. These berthing ports have been used over and over again, and I'm amazed that so far none of them has failed or shown any signs of wear. Everything has gone just as planned, and we now have 4,300 pounds of cargo to unload.

Our care packages are clearly marked and easily accessible once we open the hatch, as are the mice, the fresh food, and ice cream. Terry and I distribute the packages to everyone, feeling a bit like Santa Claus.

These items were gathered from our families and friends months ago in order to be packed into the Dragon. Care package items need to be small, light, and nonperishable. I leave mine in my CQ to open in private later.

The fresh food bags contain apples, pears, red and green peppers. They smell great. We will eat them at nearly every meal for the next few days before they spoil.

I unpack the live mice and transfer them, one by one, from the habitat they launched in to their larger, more comfortable facility in the U.S. lab. They scramble around, trying to make sense of weightlessness. I watch their faces and wonder if their tiny brains can process the change they have experienced. Like people, they're not looking too good at first.

All the cargo we unload from Dragon must be packed into labeled fabric bags. The labels have bar codes, just like food in a grocery store, as well as printed text indicating what's in each. Everything has a purpose and a destination—not only to go to a certain module, but to go in a specific bag or locker on a specific wall (or floor or ceiling) of that module. It's so easy to lose stuff up here that if we were to put something in the wrong place, we may never see it again. This makes the work of unpacking Dragon both tedious and stressful, a combination that seems to occur a lot on the International Space Station. After spending a few hours at the interface between Dragon and ISS, I notice that my arms smell like space.

SINCE IT'S SATURDAY, I have a bit more time to make personal phone calls to friends and family. I've found myself thinking about my mother today—it's been three years since she died, and though I'm not usually especially attuned to dates and anniversaries, I'm wishing she could see what I'm doing up here. She was so proud of Mark and me when we became astronauts, and she came to all six of our launches in Florida. The further I've gone in my career, the clearer it seems to me that the early lessons she taught Mark and me by example have made a huge difference in my life. Seeing her set herself an incredibly tough goal—to pass the men's physical fitness test to join the police department—and then to conquer it, was worth more than all the pep talks in the world. I remember watching her post her workout schedule on the fridge, detailing which days she would lift how much weight or how far she would run. As the weeks went by and more of those workouts were crossed out, we saw her get stronger. Her accomplishment wasn't meant to be instructional for Mark and me, but it was.

All the stories I've heard about my mother's years on the police force made me believe she was the best kind of cop. She truly cared about the people she interacted with, even if they were doing dumb stuff, and she put their safety ahead of her own. She could often defuse a situation by listening rather than threatening, and she made compassionate judgments when it might have been easier to arrest an offender. She hated to take people to jail and would often come home late because she had driven someone home herself instead. My mother sustained a lot of injuries on the job. After ten years, her back ailments had become serious enough that she retired on a disability pension and didn't work again. She wasn't sorry to leave; she had found police work very demanding, though she was proud of her service and we were proud of her. She was happy to fill her time with her artwork and, later, her grandchildren.

When I get a chance to float into my CQ, I see that Amiko has emailed me. She put some flowers on my mother's grave today and took a picture to include in the email. Seeing my mother's name on her gravestone, the bright colors of the flowers, the green of the grass all around—I'm pulled back to Earth all at once. The image reminds me both of the simple wonder of things like flowers and grass and also of the fact that we have to lose the people we love most. Most of all, I'm moved by Amiko's gesture. She has a lot to deal with on the weekends, but she remembered this date and drove out to the cemetery to do this because I couldn't.

"Thanks for doing that," I tell her when I call. "That means a lot to me." There was more I wanted to say but I can't quite put it into words. Amiko was with me at the end of my mother's life. She was with me when I learned I would be going on this mission. She knows, more than anyone but my brother, what it would mean to my mother to see what I'm doing now.

"I remembered what I wanted to tell you," I say. "I spent all day getting into SpaceX, and now my arms smell like space."

"That's so cool," Amiko says. "Tell me what it smells like." She knows, because I've told her before. She listens again.

We continue to unpack Dragon on Sunday. I work through a few bags of medical supplies, clothes, and food. I'm taking a break to do some cleaning—it's still Sunday, after all—and not long afterward I hear a fire alarm.

Astronauts do not scare easily, and this alarm does not exactly scare me, but it certainly gets my attention. Fire is on the short list of things that can kill you in space incredibly quickly. A fire on the old Russian space station *Mir* blinded and choked the crew within seconds, and if it hadn't been for their quick reaction they could have died. Some of the older cosmonauts, including Gennady, refuse to cut their hair in space, because Sasha Kaleri was cutting his hair when the fire started on *Mir*. I know as I hear the first peals of the alarm that I have set it off myself—I'm in the middle of cleaning an air filter, setting free some dust that must have triggered the sensitive smoke detector. Still, an alarm is an alarm, and everyone has to respond according to the checklist. It takes the ground quite a while to recover from the ventilation shutdown that is the automated response to any fire alarm. By the time it's resolved, I'm in a pretty crappy mood.

ON A MONDAY MORNING a couple of weeks later, I prepare to start working on the rodent experiment, designed to study the negative effects of spaceflight on mammal physiology, with the goal of developing ways to prevent them. So much of the damage to our tissues mirrors the effects of aging—muscle wasting, bone loss, cardiovascular weakening; the solutions that come from these studies will have wide-ranging benefits for humanity, not just for space.

I get out the scalpels, hemostats, tweezers, scissors, probes, syringes filled with sedatives to spare the mice pain, and fixatives to preserve the tissues. I set up the glove box, a glass case with gloves built into the front that lets me manipulate what's in the box without exposing it to the rest of the station. On Earth the glove box wouldn't be necessary for this kind of work, but in weightlessness it's worth avoiding having scalpels and all the rest floating around the lab. I've already discovered that the habitat isn't fail-safe, and I have to constantly look out for small brown UFOs while floating through the lab.

The scientists who design the experiments conducted on station try to minimize the time and attention required from astronauts. There are many experiments going on here that I know absolutely nothing about because they will be worked on by other crew members or because they don't require human involvement at all, humming along on their own either inside or outside ISS for the entire year. There are others that require only that I push a button or load a new sample once in a while. Some will take more of my time, such as this rodent experiment. I will spend all day with the mice, and the work will be precise and demanding. In order to carry it out, I trained with the scientists in charge of the research before my launch, learning the skills of dissection.

Terry takes the first mouse out of its habitat and slips it into a small container to transfer to the glove box. With nothing to hold on to, she does a slow circle in zero g, working her paws uselessly. I've been watching the mice float and wriggle, and it seems as though they have adapted quite a bit and are learning how to get around in this new environment with its new laws of physics. Even their physical condition has appeared to improve since they've been here. I set up the cameras for a live downlink with the scientists on the ground in Alabama and California who will talk me through my work in real time. I put the mouse against a piece of wire mesh—since they seem to like to have something to hold on to—then grip the loose skin on the back of her neck, the way you'd scruff a cat, hold the tail between my pinky and ring finger, and slip the needle and syringe full of sedative into her belly.

Once the drug takes effect, Terry puts the mouse into a small X-ray machine. Next I slice into her abdomen, exposing her organs, then insert a syringe into her heart and draw blood into a tiny tube, not unlike the blood I take from myself for the human research studies, except that this step euthanizes the mouse. I put the tube into a bag and label it carefully. Next I remove the mouse's left eye, following instructions from the ground. That goes into a container and gets labeled. Then I remove her hind legs. This set of experiments is specifically designed to learn about eye damage, bone loss, and muscle wasting. It's not lost on me that all of the biological processes affecting this mouse are also affecting my own body.

Early in my career as an astronaut I was skeptical of whether I wanted to fly on the International Space Station: most of what station astronauts do *is* science. After all, I'm a pilot. The goal that had driven me to become an astronaut was to fly more and more challenging aircraft until I got to the hardest thing there was to fly: the space shuttle. Dissecting a mouse is a far cry from landing a space shuttle—but then again, so is unpacking cargo, repairing an air conditioner, or learning to speak Russian, and I do those things too. I've come to appreciate that this job has challenged me to do not just one hard thing, but many hard things.

MORE THAN four hundred experiments will take place on ISS during this expedition, designed by scientists from many countries and representing many fields of study. Most of the experiments in one way or another study the effects of gravity. Pretty much everything we know about the world around us is influenced by gravity, but when you can remove that element from the subject of an experiment—whether it's a mouse, a lettuce plant, a fluid, or a flame—you have unlocked a whole new variable. This is why the science taking place on station is so far ranging; there are few branches of science that can't benefit from learning more about how gravity affects their subjects.

NASA scientists talk about the research taking place on station as falling into two large categories. The first comprises studies that might benefit life on Earth. These include research on the properties of chemicals that could be used in new drugs, combustion studies that are unlocking new ways to get more efficiency out of the fuel we burn, and the development of new materials. The second large category has to do with solving problems for future space exploration: testing new life support equipment, solving technical problems of spaceflight, studying new ways of handling the demands of the human body in space. All the experiments of which I am the main subject fall into this second category: the study comparing Mark and me as twins over the course of the year; the study on the effects of a year in space on Misha and me; the work being done on my eyes and heart and blood vessels. My sleep is being studied, as is my nutrition. My DNA will be analyzed to better understand the effects of spaceflight at a genetic level. Some of the studies being conducted on me are psychological and social: What are the effects of long-term isolation and confinement?

Science takes up about a third of my time, human studies about three-quarters of that. I must take blood samples from myself and my crewmates for analysis back on Earth, and I keep a log of everything from what I eat to my mood. I test my reaction skills at various points throughout the day. I take ultrasounds of blood vessels, my heart, my eyes, and my muscles. Later in this mission, I will take part in an experiment called Fluid Shifts, using a device that sucks the blood down to the lower half of my body, where gravity normally keeps it. This will test a leading theory about why spaceflight causes damage to some astronauts' vision.

In fact, there is much crossover between these categories of research. If we can learn how to counteract the devastating impact of bone loss in microgravity, the solutions may well be applied to osteoporosis and other bone diseases. If we can learn how to keep our hearts healthy in space, that knowledge will be useful for heart health on Earth. The effects of living in space look a lot like those of aging, which affect us all. The lettuce we will grow later in the year is a study for future space travel—astronauts on their way to Mars will have no fresh food but what they can grow—but it is also teaching us more about growing food efficiently on Earth. The closed water system developed for the ISS, where we process our urine into clean water, is crucial for getting to Mars, but it also has promising implications for treating water on Earth, especially in places where clean water is scarce. This overlapping of scientific goals isn't new—when Captain Cook traveled the Pacific it was for the purpose of exploration, but the scientists traveling with him picked up plants along the way and revolutionized the field of botany. Was the purpose of Cook's expedition scientific or exploratory? Does it matter, ultimately? It will be remembered for both, and I hope the same is true of my time on the space station.

By the end of the day working with the mice, I have a collection of sample bags the scientists on the ground are itching to get their hands on. They will have to wait until we send Dragon back to Earth, but they couldn't be more pleased with how the dissection went. Terry puts the samples in the freezer. I'm exhausted from the extreme focus and from being locked in one position all day with my hands in the glove box. But it's satisfying to know my work will be useful. I clean up, putting all the tools and instruments back where they belong, remembering that a tool in the wrong place is no better than a tool we don't have. I head into Node 1 to find some dinner. We don't go out of our way to eat together, except on Fridays, because our schedules are just too crazy to allow it. I warm some irradiated meat, douse it in hot sauce, and eat it on a tortilla, floating alone while watching an episode of *Comedians in Cars Getting Coffee*. While I'm finishing up, Terry comes by.

"Hey, don't forget we got ice cream on SpaceX," he reminds me. He goes to the tiny freezer in the ceiling of the lab and brings back a Klondike bar for each of us. It's real ice cream, not the freeze-dried stuff that's marketed as astronaut ice cream, which we don't actually have in space. I've never had ice cream in space before—we usually don't get to eat anything cold. It tastes amazing.

Back in my CQ, I look through my care package that came up on Dragon again. There is a poem and some chocolates from Amiko (she knows I crave chocolate when I'm in space, though on Earth I don't have much of a sweet tooth); a bottle of Frank's hot sauce; a postcard from Mark showing twin redheaded little boys giving the finger to the camera; and a card from Charlotte and Samantha, their distinctive styles of handwriting gouged into the heavy paper by a black pen.

I eat a piece of the chocolate and put everything else away. I check my email again. I float in my sleeping bag for a while, thinking about my kids, wondering how they are doing with me being gone. Then I drift off to sleep.

6

AT FIVE in the morning, when it was still fully dark, I slipped into the B Company dorm. I quietly opened the door of a room on the third floor, where two eighteen-year-old boys, Maritime freshmen, were sleeping soundly. The room smelled of unwashed socks and sweat. I stood over the boy on the left side, the bed I had slept in myself only two years earlier. On the other side of the room, another indoctrination officer stood over the bed where Bob Kelman had slept. When I gave the signal, we both started banging the garbage pail lids together while screaming, "Wake up, MUGs! Wake up, you lazy bastards!" at the top of our lungs.

I had been appointed the chief indoctrination officer for the class, in charge of supervising all those running the grueling period of drills and training the incoming freshmen. It was a demanding job but a huge honor—it meant I had done exceptionally well and that my superiors saw leadership potential in me. I was determined to prove them right. This was my first real opportunity to be a leader.

I had 250 new MUGs (Midshipmen Under Guidance) to train. I was responsible for teaching them the traditions and expectations of Maritime, as well as helping them adjust to life away from home. As the final authority on discipline, I had decided that I wanted to be the kind of leader who was firm but fair. I wanted to hold everyone to the same high standard, but I also wanted to approach each situation with an open mind and a willingness to listen to others' points of view.

I once received an anonymous note from a MUG warning me not to get too close to the ship's railing at night on our next cruise—a threat to push me overboard. This was an early lesson that a leader can't always please everyone. I can understand why this MUG and the others I dealt with found the rules burdensome. But I had come to believe that shined shoes and polished belt buckles, however insignificant they might seem, helped us to learn the attention to detail required to safely and effectively operate at sea.

Each summer, we took the *Empire State V* to new ports, and immediately after I returned from each of those cruises I would then leave for my Navy cruise. I spent one summer doing a program called CORTRAMID (Career Orientation and Training for Midshipmen). We spent a week each in the surface, submarine, and aviation communities as well as a week with the Marine Corps. The idea was to give us some exposure to the different options for Navy service. With the Marine Corps, I observed explosives demonstrations and ran around in the woods with an M16 at night. With the aviators I flew in an E-2C Hawkeye aircraft, and with the Navy SEALs I got to do their grueling obstacle course. I spent three days on a submarine.

In my senior year I was named the battalion commander of my Navy ROTC unit, another leadership role. By that time, I was taking harder classes than ever, mostly electrical engineering. I now knew how to study and took pride in it, actually enjoyed it. I was learning circuit design, network analysis, and other advanced engineering courses. I would have liked to change my major to physics if that had been an option at Maritime. I've sometimes thought if I were ever to become a college professor, I would want to teach first-year physics or calculus. Those foundation classes are make-or-break for students, and I think it would be rewarding to give young people the keys to learning hard things that I had figured out for myself.

It was still my goal to become a Navy pilot, specifically to fly jets off an aircraft carrier. In college, I had been doing whatever I could to improve my chances, including caring for my vision. A lot of my friends who hoped to become pilots talked about how to maintain their vision, and we all became a bit obsessed. Every prospective pilot knew some poor bastard who had worked all his life toward becoming a Navy pilot only to be rejected for having vision slightly less than 20/20. I was concerned about eyestrain and made sure to always have a bright light to read by. In retrospect, there was probably nothing I could have done to have much effect.

Early in my senior year, I took a standardized test called the Aviation Qualification Test/Flight Aptitude Rating. The qualification test was something like an IQ test, and the flight aptitude part consisted of mechanically oriented puzzles and a visual logic section that showed illustrations of views of the horizon from a plane's cockpit that we had to match with the correct airplane orientation.

I knew how important this test would be to my future, so I worked hard to prepare for it. There weren't study guides, so I made my own, drawing pictures of airplanes and what the view would look like from the cockpit. The day of the exam, I left the classroom feeling like I had done as well as I possibly could. I wouldn't know for weeks what my results were, and then it would be months after that before I would learn to what part of the Navy I would be assigned. Even if I did well, there was still no guarantee that I would be chosen for aviation, much less that I would go on to fly jets.

ONE COLD DAY in January, my roommate George Lang and I were sitting in our room just after lunch, watching *Star Trek* on the tiny color box TV we kept next to the fish tank in our room. A news anchor broke into the show to report that the space shuttle *Challenger* had exploded seventy-three seconds after launch. We watched the shuttle blow up on the screen over and over, just after the ground gave the call "go at throttle up." (At the time I had no idea what this phrase meant; much later I would learn to respond to it myself, confirming the communications between the ground and the shuttle.) It would be weeks after the accident before the theory emerged that the unusually cold weather in Florida had caused a rubber O-ring in one of the solid rocket boosters to fail.

"You still want to do it?" George asked me after a few hours of watching nonstop.

"What do you mean?" I asked.

"The shuttle," George said. "You still want to fly on it?"

"Absolutely," I said, and I meant it. My determination to fly difficult aircraft had only grown stronger as I had learned more about aviation, and the space shuttle was the most difficult aircraft (and spacecraft) of all. The *Challenger* disaster had made clear that spaceflight was dangerous, but I already knew that. I felt confident that NASA would find the cause of the explosion, that it would be fixed, and that the space shuttle would be a better vehicle as a result. It sounds strange, but seeing the risk involved only made the prospect of flying in space more appealing.

It wasn't until years later that I understood that a management failure doomed *Challenger* as much as the O-ring failure. Engineers working on the solid rocket boosters had raised concerns multiple times about the performance of the O-rings in cold weather. In a teleconference the night before *Challenger*'s launch, they had desperately tried to talk NASA managers into delaying the mission until the weather got warmer. Those engineers' recommendations were not only ignored, they were left out of reports sent to the higher-level managers who made the final decision about whether or not to launch. They knew nothing about the O-ring problems or the engineers' warnings, and neither did the astronauts who were risking their lives. The presidential commission that investigated the disaster recommended fixes to the solid rocket boosters, but more important, they recommended broad changes to the decision-making process at NASA, recommendations that changed the culture at NASA—at least for a while.

Years later, one of the first briefings I got as a new astronaut was about the *Challenger* disaster. Hoot Gibson, who was in the same class as three of the *Challenger* crew, detailed exactly what had gone wrong that January day. He also told us what the crew likely experienced in the last minutes of their lives. He wanted us to understand the risks we would be running if we flew in space. We took his words seriously, but no one dropped out after that briefing.

. . .

GRADUATING FROM MARITIME, in 1987, made me pause and reflect. My admission had been make-or-break for me. I would never forget that. What I had learned there—in the classroom, on the ship, from my peers, from my mentors—had been life changing. I was a completely different person from the confused kid who had entered through those gates four years earlier. I felt a debt of gratitude to the school for everything it had done for me, and I was nostalgic about leaving a place where I had so many fond memories. Over the years, I've tried to stay connected to the school, and in the time since I graduated their prestige has grown—when financial magazines rank colleges whose graduates have the highest salaries, SUNY Maritime is almost always up there with Harvard and MIT, sometimes at the very top.

I earned a high score on the aviation qualification test, and soon after I was assigned to flight school in Pensacola, Florida. I packed all my belongings into my old white BMW and drove south that summer of 1987. Pensacola is on the panhandle, commonly known as the Redneck Riviera, so in a lot of ways it's more like Alabama than most people's idea of Florida. It's a small city, dominated by the naval air station, and tourism is the main industry aside from training Navy fliers. Pensacola is very much a typical military town with trailer parks, pawn shops, and liquor stores, but in this case set against a background of beautiful beaches.

When I reported for my eye exam on the first day of flight school, there were four uniformed officers facing me. I'd expected to find one busy flight surgeon who would make me read a chart and then (I hoped) send me on my way, but the wall of high-ranking officers scrutinized me, stern and unsmiling throughout the exam. Their presence was distracting, and I continually questioned my responses—perhaps this was their intention. I got through the eye exam with a clean bill of ophthalmic health. Years later, I met a Navy flight surgeon who was in that room the day of my eye exam. He admitted that it was an intentional tactic of intimidation.

Naval aviation indoctrination got started with several weeks of tough physical, swim, and survival training. There was a cross-country course we had to navigate in a certain amount of time, an obstacle course with hurdles to jump over, barriers to shimmy under, sand to crawl through, a wall to climb. The film *An Officer and a Gentleman* gives a pretty accurate representation of what aviation indoctrination training was like, and just as in the film, we student naval aviators had to conquer the Dilbert Dunker a few weeks in. The dunker is designed to simulate the unpleasant experience of a water landing or ditching in an airplane. Dressed in full flight gear and helmet, we were strapped into a mock-up cockpit that was then sent down a steep rail into the deep end of a swimming pool. We were warned that the impact with the water could be hard enough to knock the wind out of us, and that once submerged we'd have only a few seconds to get our bearings before the cockpit turned upside down. I would have to detach the comm wire from my helmet, release myself from the restraints from which I'd be hanging, find my way out of the cockpit, and then dive deeper in order to escape the fuel that might be burning on the ocean surface in a real water landing. A few people who went through this before me couldn't find their way out and had to be pulled out of the cockpit by rescue divers. This made the risks much more vivid to those of us still standing in line, but when I hit the water I managed to find my way out on the first try.

We also had to go through a similar dunker that simulated a helicopter crash in water. We were strapped into a mock-up helicopter, which was dropped into a pool, flipped over, and sent to the bottom. As with the Dilbert Dunker, I had to be able to get unstrapped and swim to safety. The helo dunker was much harder, though, because several of us, blindfolded, had to get out a single door. People have drowned in the helo dunker, and I heard that some even went into cardiac arrest. We sat strapped in and watched the water slowly climb, grabbing a last breath as it reached our noses. We had to wait to unstrap ourselves until after we were upside down and the motion stopped. I'd try to find a railing or structure on the inside of the cockpit to serve as a reference point to reach for once blindfolded. Once I was upside down, though, everything seemed to move around, and I'd inevitably get kicked in the face by someone flailing for the door, or get kicked in the stomach and have the wind knocked out of me. I'm sure I also kicked the guys behind me. I couldn't have been happier when I passed the test, though I knew I would have to requalify every four years (NASA has its own water survival training, but it's much easier). As it happens, I would never need to use any of the emergency training, either the Navy's or NASA's.

The swim requirements were even harder. We had to be able to swim a mile and tread water for fifteen minutes, in full flight suit and boots. I got through the mile easily, but I found treading water murderously difficult. Other guys seemed to be naturally buoyant; I seem to have the buoyancy of a brick. I practiced and practiced and was finally able to pass the requirement, though just barely.

I also learned various survival techniques in water, like taking off my pants and making a flotation device out of them by tying the legs closed and filling them with air. I learned drownproofing, a technique for staying alive in water for long periods of time by calmly floating facedown in the water and bringing my mouth slowly up to the surface only when I needed to take a breath. I learned how to disentangle myself from the strings of a parachute collapsed on me in the water. I practiced being rescued from the water by a helicopter, hooking a sling called a horse collar around myself to be hoisted up into the air. The hardest part of this was all the water the helo would kick up into my face, making it feel as though I was drowning.

One day we were taken in groups to experience the altitude chamber, a sealed room in which the air pressure is slowly lowered to simulate an altitude of 25,000 feet. At this level the oxygen deprivation isn't life threatening, but it gave us a chance to observe our symptoms of hypoxia, which can include tingling in the extremities, nails and lips turning blue, trouble speaking clearly, and confusion. After a number of sessions in the chamber, I tried to push my limits to see how bad my symptoms could get. At first I started to feel a bit drunk and stupid, a vaguely pleasant sensation that turned quickly into euphoria. Euphoria became confusion, followed closely by tunnel vision, and the next thing I knew, the safety monitor was putting my oxygen mask back on for me—I had waited too long and become unable to do it myself. The lesson of the low-pressure chamber was that you go over the cliff quickly. I would continue to do periodic recertification in the altitude chamber, but I always avoided the cliff.

We also did a great deal of coursework. We learned aerodynamics, flight physiology, aircraft engines and systems, aviation weather, navigation, and flight rules and regulations. Most of this material was new to me, but it wasn't too dissimilar from what I had studied in college. Some of my classmates who had chosen undergraduate majors in the arts and humanities struggled more with the material. But I knew this was one aspect of the training I could excel at if I applied myself to it, so I did. The grades we earned didn't count in the same way a GPA does in college, but I knew that the better I did at every aspect of aviation indoctrination the better my chances would be of getting assigned to jets.

As part of our survival training, we were dropped off in the woods for days to learn to build shelters, make signal fires, navigate on land, and feed ourselves on only what we could hunt or forage. We couldn't find anything to eat except for a rattlesnake we killed with a big stick.

PENSACOLA WAS the top of the world for a young officer like me earning a salary for the first time—the princely sum of $15,000—with no dependents and no responsibilities other than to the Navy. I walked around town feeling like a rock star and spent a lot of that salary in bars. At Trader Jon's, a dimly lit dive, the brick walls were crammed with photos of pilots and other aviation memorabilia, and metal model planes hung precariously overhead. At a bar called McGuire's, hundreds of thousands of one-dollar bills signed by the patrons dangled from the ceiling like sleeping bats. I added one of my own.

After we got through classroom and physical training successfully, which took about six weeks, it was finally time to learn to fly airplanes. We started off flying the T-34C Turbo Mentor, a propeller-driven trainer. It's a post–World War II–era airplane, small, with a tandem seating arrangement, one seat in front and the second in back. The flight manuals we had to study were phone-book thick, packed with charts and graphs and studded with unfamiliar terms and abbreviations. The material was incredibly dry, but we had to master it before we could fly.

My strategy was to study everything assigned for each day and get ahead on the next lesson's reading as well. I committed the emergency procedures to memory as I'd been told to. If the instructor asked me what I would do if I lost an engine on the T-34, I could tell him, "Put the PCL idle, T-handle down clip in place, standby fuel pump on, starter on, monitor N1 and ITT for start indications, starter off when ITT peaks or no indication of start." I haven't flown the T-34 for nearly thirty years, and I only flew it a total of seventy hours, but I can *still* rattle this off without thinking. I could still recover from the loss of an engine, or a range of other emergencies, in that plane.

When I was declared ready, the first phase of my actual flight training began. In the briefing room, I met Lieutenant Lex Lauletta, my on-wing instructor, a tall blond guy who greeted me with a congenial smile. That set me at ease, since some of the instructors were said to be real assholes, especially to guys like me who were dead set on flying jets. Lauletta was a former P-3 pilot who was building his flight hours in order to become an airline pilot. I would do most of my initial flights with him, and he kept me from killing myself as well as instructing and mentoring me. He would also be grading me, and his evaluation would count more than anything else to determine whether I would get to go on to meet my goals of flying jet aircraft or would be sent to fly helicopters or larger fixed-wing airplanes—or nothing at all.

That day in the briefing room, we talked about what the syllabus looked like, what we would do when we met next, and how my preparation was going. During that initial meeting I tried my own "green bag," or flight suit, for the first time. For me, this was like getting assigned a uniform you get to wear for the rest of your flying life that lets people know you're a badass Navy pilot. I would rarely go to work wearing anything other than a flight suit for the next nine years.

Later, we walked out to the airplane for the first time. It was a cold, foggy fall morning, weather I wouldn't be allowed to fly in alone for a long time. As I got strapped in, I was excited and nervous. I had invested so much in the idea of being a carrier aviator, had worked so hard to get to this point, but I had no idea if I could actually fly a plane. Some people can't, no matter how hard they try, and you can't know that until you're up in the air.

Out on the airfield, I saw hundreds of T-34s were lined up, one after another, stretching out into the horizon, their distinctive bubble canopies covered with condensation. Lieutenant Lauletta figured out which airplane was ours, and as we walked toward it he gave me my first lesson about how not to get killed: never walk through a propeller arc, even if you know the propeller isn't turning. When he found the airplane that had been assigned to us, he jumped up on the wing, opened both canopies, and threw our helmet bags onto our seats—his in back and mine in front.

He led me through my first preflight check. We checked the wings, flaps, and flight control surfaces on the wings, then opened the engine cowling and inspected the engine, including checking the oil. We looked at the propeller, checking for damage. We checked that the tires were properly inflated and that the brake pads weren't overly worn. We agreed that everything looked normal, though in reality I wouldn't have been able to tell if something was wrong. Lieutenant Lauletta tried to give me as much detail as he could about what he was looking for. Then it was time to climb into the plane.

The first moment I settled down into the seat was surreal. On one hand, it was the end of a long struggle to get there, starting the afternoon I first cracked the cover of *The Right Stuff*. There had been many moments when it seemed that I wasn't going to make it. Now I could say I had—I was a student naval aviator. On the other hand, this was going to be the start of a whole new set of challenges.

Lauletta helped me get strapped in properly, then we both closed our canopies. I'd studied diagrams of the cockpit of the T-34 in the flight manual as if my life depended on understanding them (because it would). I'd learned the controls and practiced using them in the simulator. Now they seemed to have multiplied into a field of thousands of knobs, switches, gauges, and handles. I had to tell myself to get on with it, that I was ready to do this. It was time to start the plane.

Under Lauletta's instruction, I applied power and started moving forward. Taxiing was more difficult than I had anticipated, because the airplane didn't have nose-wheel steering, like a car. Instead, I had to use differential braking to steer the plane, meaning I would partially apply the brakes just on the left side if I wanted to turn left, and just on the right side if I wanted to turn right. This was so completely counterintuitive, I felt like I was learning to ride a bike, trying to keep my balance with someone watching over my shoulder the whole time, grading me. I was already struggling.

A pilot must also learn to use the radio, which is harder than you would think. Talking and doing anything else at the same time can be challenging, as it requires using two different parts of the brain. And of course I wanted a cool Navy radio voice. When Lauletta cued me, I spoke into the radio and said, "Whiting tower, Red Knight Four Seven One ready for takeoff."

Somehow this did not sound nearly cool enough to me. I felt like a little kid playing make-believe. But the tower responded as if my call had been legitimate. "Roger, Red Knight Four Seven One, taxi to position and hold." This meant we could head out onto the runway but weren't cleared for takeoff yet. Eventually the tower came back: "Red Knight Four Seven One, you are cleared for takeoff."

I ran the power all the way up to maximum and accelerated down the runway, trying my hardest to keep the airplane pointed in the right direction using the toe brakes. Once I was going faster, it was a bit easier to control the plane's direction using the rudder, and with Lauletta's instruction I slowly pulled back on the stick to make the nose come off the ground. The runway, buildings, and trees tilted back and fell away as we pointed up into the sky. We porpoised a little, undulating up and down as I struggled to find the proper attitude—an aircraft's orientation in the sky—but we were airborne. In that moment, I was elated. I was flying a plane—albeit very poorly.

We headed out using the "course rules," a set of formal instructions on where to fly using reference points on the ground. These rules are designed to keep student naval aviators from crashing into one another in the air. I checked in on the radio, announcing where we were so other pilots could avoid us.

Once I was settled into the flight, I could concentrate on mastering the most basic skill: maintaining altitude. I looked out the window at the horizon to judge my attitude, and though we were going only 120 miles per hour, I lifted and dropped us wildly, struggling to keep the airplane within five hundred feet of our intended altitude. Years later I would fly the F-14 at more than twice the speed of sound and control the space shuttle in the atmosphere many times faster, but nothing ever felt as hard to control as that training airplane on that first flight. It seemed to resist my efforts at every turn.

After about forty-five minutes of demonstrating how bad I was at this, I was relieved when Lauletta directed me toward an outlying airfield so we could practice touch-and-go landings. He demonstrated the first one, carefully describing everything he was doing. He slowed the airplane as he approached the runway, lowered the landing gear and then the flaps, came in low over the threshold of the runway, and then idled the throttles and showed me how to slow enough to touch down without stalling or losing control. He then added power and immediately got airborne again—a touch-and-go landing. He made it look easy, and in fact the T-34 is a relatively easy airplane to fly, which is why we started out on them. Now it was my turn.

Landing an airplane requires controlling the direction, altitude, and airspeed to put it down within the first few hundred feet of the runway, gently enough that the landing gear don't drive through the wings. Despite the airplane being small and the runway large, and despite the controls being relatively simple and responsive, I had a surprisingly difficult time putting landing gear and runway together properly. Eventually I managed to smack the wheels down onto the runway without killing us, then immediately took off to do it one more time, then another, then another. I didn't feel I was getting any better.

I had hoped I would fly well from the outset, but it was already clear that this was going to take some time to learn, and nothing about it was going to come easily. Still, Lauletta said I had done pretty well for my first day, and he gave me an above-average mark on "headwork," meaning that I had come prepared and that I had made good choices. This was one of the few subjective criteria he could grade me on out of ten or fifteen categories. I think he was trying to reward me for having a good attitude. He couldn't reward me for much else.

We started off flying visual flight rules, which means flying in good weather conditions so the pilot can see the horizon and avoid any obstacles or other aircraft. After twelve flights with an instructor, I was declared "safe for solo."

The first time a pilot flies solo is a big day. I climbed into the airplane not feeling particularly confident; I hadn't slept well the night before because I'd been too busy lying in bed thinking about ways I could screw up. The weather was perfect, though, with clear skies and low winds. After a good takeoff and a flight of about an hour and a half, during which I demonstrated my competence by maintaining altitude and airspeed while not crashing into anything, it was time to land. In my mind, I ran through the steps I'd performed the other times I landed. One important thing to remember was to lower the landing gear below a certain speed. I was so intent on all the things I needed to do to land the airplane that I released the landing gear too early, while I was still going fast enough that the aerodynamic forces could damage them, or in the worst case break them off. I knew I had screwed this up the second I did it, but there was no way to undo it. I had to fess up.

I called down to the tower. "Tower, Red Knight Eight Three Two."

"Go ahead, Red Knight Eight Three Two."

"I lowered the landing gear too fast, but all the gear are showing down and locked." I cringed as I waited for the response to come back.

"Okay, circle overhead at fifteen hundred feet until we figure out what we want to do. How much fuel do you have?"

I reported the fuel level, feeling relieved that the controller didn't seem very alarmed by this turn of events—he sounded just as bored by this exchange as he had been by the rest. The decision was made to have me fly by the tower so the controller could look at my landing gear and confirm that they were down and undamaged. They were, and I was allowed to land.

It's not unusual for a student pilot to commit this kind of error on a first flight, and I knew I could recover from it. Still, I was disappointed. I'd wanted to absolutely nail everything the first time I soloed.

There is a saying in the Navy about mistakes: "There are those who have and those who will." It's easy to look at someone else's screwup and say, "I never would have done that." But you could have, and you still may. Bearing this in mind can guard against the kind of cockiness that gets pilots killed, and in retrospect my error overspeeding the landing gear was a good early lesson.

There was a T-34 simulator, and some of our graded flights were "flown" in the simulator rather than in the airplane. We could sign up for practice time in the simulators, and whenever the new schedule went up, I was first in line to book as much time as I was allowed. I did extremely well on all my graded simulator flights, and since the flight instructors had to help us set up the simulator for the practice sessions, it didn't hurt that I was making an impression on them as a motivated student.

Once I had soloed a few times, I started learning aerobatics. I went out with an instructor again, listening while he explained the maneuver he was about to demonstrate. I found I had a real knack for it, and I enjoyed this part of the training—the sense of freedom it gave me—more than anything else. Flying around the big, puffy clouds, rolling the airplane upside down and around at will, feeling the force of acceleration pushing me down into my seat—I never felt like I was disoriented or sick, which happened to some of the other newbie pilots. It felt great to find an aspect of flying I was good at. As I finished up that part of the syllabus I couldn't wait to try aerobatics in a more powerful plane, and I couldn't wait to fly that way while simultaneously pretending to shoot another airplane out of the sky.

Some people washed out even before they got the chance to fly solo: they couldn't pass the swim requirements, couldn't pass the survival training, or failed their safe-for-solo check flight. The program wasn't meant to weed people out—the Navy had already invested a lot in each of us, and they wanted us to succeed—at the same time, they needed to be sure we wouldn't endanger ourselves or others. Only a small percentage of those who start flight school wind up being assigned to a jet squadron, and I had done everything I could to establish a place among them.

We knew that our next assignments would be announced on an upcoming Friday. That day, we waited in the hallway to learn our fate. I didn't feel as nervous as some of my classmates seemed to be. I knew I had made every effort and held nothing back, working as hard as I could at the things I could control and ignoring what I couldn't. I was ready for whatever was to come.

Finally a secretary tacked a simple sheet of paper to the bulletin board. We all crowded around. It had ten names on it in alphabetical order, and next to each one an assignment. Next to KELLY, SCOTT I found the words BEEVILLE NAVAL AIR STATION. I had done it. I was one of two guys in my group to make it to jets. I felt for my friends who didn't, but I was elated knowing that my dream was still alive.

The International Space Station (ISS) with its giant solar arrays appearing to cut through the blackness of space as the moon beckons in the background

The ISS seen with Earth below

Mark, left, and me circa 1967 in our yard on Mitchell Street, West Orange, New Jersey

My father is holding the Bible as my mother is sworn in as the first female police officer in West Orange, August 1979.

My Radio Intercept Officer, Bill "Smoke" Mnich, and me flying an F-14 Tomcat in 1995. My years as a test pilot preceded my years as an astronaut.

My astronaut class photo, 1996. We were known as the Sardines. We were all from the U.S. unless otherwise indicated below. From left to right and back to front: Christer Fuglesang (Sweden), John Herrington, Steve MacLean (Canada), Peggy Whitson, Steve Frick, Duane Carey, Dan Tani, Heidemarie Stefanyshyn-Piper, Jeff Williams, Don Pettit, Philippe Perrin (France), Dan Burbank, Mike Massimino, Lee Morin, Piers Sellers (deceased), John Phillips, Rick Mastracchio, Christopher Loria, Paul Lockhart, Charlie Hobaugh, Willie McCool (deceased), Pedro Duque (Spain), Soichi Noguchi (Japan), Mamoru Mohri (Japan), Gerhard Thiele (Germany), Mark Polansky, Sandy Magnus, Paul Richards, Yvonne Cagle, Jim Kelly, Pat Forrester, Dave Brown (deceased), Umberto Guidoni (Italy), Mike Fincke, Stephanie Wilson, Julie Payette (Canada), Lisa Nowak, Frank Caldeiro (deceased), Mark, Laurel Clark (deceased), Rex Walheim, me, Joan Higginbotham, and Charlie Camarda

My space shuttle *Discovery* crew, headed to the launchpad for my first spaceflight. From left to right, front to back: me, Curt Brown, John Grunsfeld, Jean-François Clervoy (France), Mike Foale, Claude Nicollier (Switzerland), and Steve Smith

Lifting off in *Discovery,* December 1999

On my first mission to space, we repaired the Hubble Space Telescope and sent it on its way to continue the exploration of the universe, Christmas Day, 1999.

Left to right: me, Dima Kondratyev, and Sasha Kaleri awaiting water survival training on the deck of a Russian naval vessel in the Black Sea, September 11, 2001. I learned of the disaster when the ship returned to port.

Donning my harness prior to entering space shuttle *Endeavour* as the commander of STS-118, August 2007

My STS-118 crew and the crew of the space station in the U.S. laboratory module. From left to right and front to back: Clay Anderson, Fyodor Yurchikhin (Russia), Oleg Kotov (Russia), Al Drew, Barbara Morgan, Dave Williams (Canada), me, Charlie Hobaugh, Rick Mastracchio, and Tracy Caldwell

Shannon Walker, Doug Wheelock, and Fyodor checking out their spacesuits prior to returning to Earth from Expedition 25, my first long-duration flight on the ISS, November 2010

Hanging out of our crew quarters in Node 2 of the ISS. From left: Doug, me on the "ceiling," Shannon, and Oleg in the "floor"

My daughters, Samantha, top, and Charlotte, in Red Square, Moscow, in the summer of 2008

My partner, Amiko, and me in Red Square

In my Star City, Russia, cottage in February 2015, studying for my final exams before launching to the ISS to spend a year in space

7

April 25, 2015

Dreamed I was on Earth. I was floating a few feet above the ground, to be exact, flying around New York City. I flew over the George Washington Bridge, down Fifth Avenue, through the Holland Tunnel, over to New Jersey, and around Giants Stadium. No one seemed to notice me. I was doing something important as I was floating around, maybe some kind of antiterrorism reconnaissance.

TODAY IS SATURDAY, almost two months into my mission, and Terry is euthanizing a mouse. Last night we got a call from the ground telling us that one of the mice was "in distress" and would need to be put down today. When we look into the cage first thing in the morning, we find the distressed mouse in a terrible state—missing a limb, apparently chewed off by the other mice or by herself. We work quickly to give her an injection. We are upset to know the mouse was suffering all night while we were sleeping. We tell mission control that in the future we want to know about a situation like this right away. They were trying to protect our time, but we would have liked to make that choice for ourselves. They seem surprised by how strongly we feel about it. I haven't made the mistake of getting too attached to the mice, knowing what their fate will be. But it's been hard not to take an interest in them, as their bodies go through the same changes ours have. They started off looking sick and disoriented, moving awkwardly, but as the days go on they look healthier and get better at negotiating the subtleties of moving around in zero g, just as we do.

When we got the call about the mouse last night, we were just finishing up with movie night—*Gravity.* We'd set up the big screen in Node 1 facing the lab and gathered to watch it—all of us but Samantha, who was finishing her workout. I've noticed a strange phenomenon when people watch movies in space: we instinctually move to a position that looks like lying down with relation to the screen. In weightlessness our positions make no difference in the way we feel physically, but the association between lying down and relaxing is so strong that I actually feel more relaxed when I get into this position. The film was great—we were impressed by how real the ISS looked, and the five of us were an unusually tough audience in that regard. It was a bit like watching a film of your own house burning while you're inside it. When Sandra Bullock got out of her space suit and floated in her underwear, Samantha happened to come floating by the screen in her workout clothes—later I regretted failing to get a picture of them together.

After we finish up with the mouse and communicate with the ground about it, I have my first videoconference with Charlotte. Unlike phone calls, these conferences have to be planned in advance. I'm ready with my laptop and my headset on at the appointed time, and when Charlotte's round face pops up on my screen she breaks out in a huge smile. The interface is similar to Skype or FaceTime—I can see Charlotte's face and the room behind her in a large window on the right of my laptop screen, while on the left is a smaller window that shows me floating in my CQ. I haven't seen her since Baikonur, a month ago. She is eleven and looks different every time I see her—she seems to have aged a year.

Charlotte is a quiet and thoughtful person by nature, and self-reliant for her age. We connect well in person, but it can be hard to have a phone conversation. During the time I've been here, I've struggled to connect with her. When I ask her, "How was school today?" I get "Fine." When I ask her, "How's your mom?" I get "Good." "How's the weather?" "Okay." She also hasn't been great about answering emails, even though she is a good writer. It's disconcerting to reach out to her and never hear back. If she was hurt or unwell or depressed, would she or someone else tell me? In our videoconference she is much more responsive and engaging, true to form. I've never been to the apartment she shares with her mother in Virginia Beach, so this is my first glance inside. I can see a small living room, a sofa, a bookshelf. In the background, Leslie walks back and forth with a laundry basket.

I spend an hour showing Charlotte around the station. She saw it in videoconferences when I was here before, but she was just seven then. I float around with my laptop, pointing its camera around inside the modules where I've been working and living, introduce her to my crewmates as they happen to float by, and give her an overview of what I've been working on (leaving out the rodent we euthanized). She seems genuinely interested, leaning forward and smiling and asking questions. It's great to see her animated and engaged. The Cupola is the last stop on our tour, and I time our arrival there to coincide with the station flying over the Bahamas. Charlotte is impressed. While we are talking I snap a few pictures with the camera to email her later. I know she's seen many pictures of the Earth taken from space, but I hope she'll enjoy receiving one taken especially for her.

After saying good-bye to Charlotte, I start getting ready for a birthday dinner for Samantha Cristoforetti. Birthdays are important in Russian culture too, and we make a point of celebrating them up here. This one is especially significant because soon Samantha, Terry, and Anton will be leaving us. As much as I will miss them all, I'm looking forward to breathing some good air (with half as many people exhaling, CO_2 levels will come down). I know the dropping CO_2 will likely cause the ground to act as though the problem has resolved itself, and I will be upset if that happens.

When I was packing the few personal belongings I could bring for this year, I included some wrapping paper because I knew I would be giving crewmates gifts on special days. Today I have some chocolate wrapped up nicely for Samantha's thirty-eighth. As we often do at these dinners, we get to talking about language, specifically the nuances between curse words in English and Russian. Tonight we reach a point of confusion about the multiple ways of using the Russian word for "whore" and decide to call one of our Russian language instructors in Houston. Waslaw tries to explain to us in a combination of Russian and English the difference between *blyad* and *blya.* (He and I became friends on a St. Patrick's Day years ago when a Moldovan drunk started a bar fight with the NASA folks in Star City, Russia.) Then he catches us up with what's happening in Houston, and we fill him in on what life is like up here. By then it's pretty late. For a little while, it almost felt like an earthbound Saturday night. It was nice to forget for a while that I will continue to be at work up here for weeks and months and seasons.

ESSENTIAL TO getting to Mars, or anywhere else in space, is a working toilet, and ours does more than just store waste—our urine processor distills our urine into drinking water. A system like this is necessary to interplanetary missions, since bringing thousands of gallons of drinking water to Mars simply wouldn't be possible. On the International Space Station, our water system is nearly a closed loop with only occasional need for fresh water. Some of the water we purify to make oxygen.

We are sent fresh water on resupply rockets, but we don't need it often. The Russians get fresh water from the ground, which they drink and turn into pee, which they give to us to process into water. Cosmonaut urine is one of the commodities in an ongoing barter system of goods and services between the Russians and the Americans. They give us their pee, we share the electricity our solar cells have generated. They use their engines to reboost the station into the proper orbit, we help them when they are short on supplies.

Our urine processor, though, has been broken for about a week, so our urine is simply filling a holding tank. When it's full—it takes only a few days—a light will come on. In my experience, the light tends to show itself in the middle of the night. Replacing the tank is a pain in the ass, especially for a half-asleep handyman, but it's not an option to leave it for the morning. The first person to get up won't be able to pee, which isn't good space station etiquette. When you float in there in the middle of the night to find that light illuminated, it really sucks.

Now, in the light of day, I need to swap out the broken part, the distillation assembly. I've consulted with the ground, and they concur. If everything goes right, the repair will take half the day. I've removed the "kabin" (the walls and the door) from the toilet in Node 3 so I can get at the machinery underneath. (The spelling is attributed to a transliteration error between Russian and English that stuck.) The kabin gets pretty gross, even though we try to clean it regularly. I float the kabin to Node 1, where it will clog up this space for other people until I move it back again, another incentive to get the job done efficiently.

While I'm cleaning and then moving the kabin, the ground is taking care of "safing" the equipment, which means making sure everything I will be working on is powered down correctly so I don't electrocute myself or cause an electrical short. (The risk of electrocution is ever present on the space station, especially on the U.S. side. We use 120-volt power, which is more dangerous than the 28 volts used on the Russian segment. We train for the possibility of electrocution and often practice advanced cardiac life support on board, using a defibrillator and heart medications meant to be injected into the shinbone.) Once I get word from the ground that I can go ahead, I remove the electrical connectors on the distillation assembly, put caps on the connectors to protect them, and undo all the bolts. The distillation assembly is a large silver drum that works like a still, evaporating water from the urine. This is our only backup, so I have to be careful not to damage it.

Another resupply rocket launched today from Baikonur, a Russian Progress. My Russian crewmates on station followed the launch closely, getting updates from Russian mission control, and Anton floated down to let us know when it had reached orbit successfully. But now, less than ten minutes later, mission control in Moscow reports that a major malfunction has occurred and that the spacecraft is in a wild out-of-control spin. None of the workarounds they try correct the problem.

Up here, we talk about what it will mean for us if Progress is lost. We go over the supplies we have on board—food, clean clothes, oxygen, water, and replacement parts. Another resupply rocket exploded on the launchpad last October, this one built by the American company Orbital ATK, which means we are already behind on supplies. The

Russians will run low on food and clothing, which means we will share ours with them and eventually run low ourselves. Misha, Gennady, and Anton keep us updated throughout the day, each time looking more and more concerned. Each of the cosmonauts had some personal items on board Progress, and sometimes those packages contain jewelry and similar irreplaceable items. Misha confides in me about some of the items that are on board, his wide blue eyes showing his anxiety.

"Maybe they'll regain control of it," I tell him with a pat on the shoulder, though we both know this is becoming less likely by the minute. I would like to spend more time talking over the problem with my crewmates, but I have a half-assembled toilet to fix. I'm disconnecting and capping the connections where our urine flows into the assembly on one end and where the liquid by-products left over, brine and a kind of graywater, come out. Every few days we pump the brine out of the holding tank and into Russian tanks that will later be pumped into empty water tanks on Progress, which will eventually undock and burn up in the Earth's atmosphere. The graywater will be processed into drinking water.

I pull out the broken distillation assembly, double-bag it, label it, then store it in the PMM (Permanent Multipurpose Module, sort of a storage closet off Node 3) until it can be returned to Earth on a SpaceX. Engineers on the ground will examine it and, if they can, repair it to be sent up again. The next step is to fit the new assembly in place and torque it to a specific value. I start hooking up the fluid lines again very carefully, making sure not to combine clean water and urine lines, then connect the electrical cables. I am taking pictures of all of my work so the ground can later verify I did everything properly.

As I'm working, the ground tells us Progress has officially been declared lost. With a sinking feeling, I float over to the Russian segment to confer. Misha meets me in the service module, and it's clear he's heard the bad news.

"We'll give you guys anything you need," I say.

"Thank you, Scott," Misha says. I don't think I've ever seen such despair on another man's face. We don't normally worry about shortages, but losing Progress suddenly makes us think about how much we depend on a steady stream of successful resupply missions. We can afford one or two failures, but then we will have to start rationing.

Even more than our concern about supplies, though, is concern for our colleagues who will be launching soon: the rocket that doomed Progress is the same rocket that launches the manned Soyuz. Our three new crewmates, due up in a little less than a month on May 26, are about to trust their lives to the same hardware and software. The Russian space agency must investigate what went wrong and make sure there won't be a recurrence. That will interfere with our schedule up here, but no one wants to fly on a Soyuz that's going to do the same thing this Progress did. It would make for a horrible death, spinning out of control in low Earth orbit knowing you will soon be dead from CO_2 asphyxiation or oxygen deprivation, after which our corpses would orbit the Earth until they burn up in the atmosphere months later.

I finish making all the connections on the urine processor. Some of the cargo that was lost on Progress was fresh water, and unless we can make our own, the six of us won't last long. I double-check all the connections, then ask the ground to power it up. It works. The ground congratulates me, and I thank them for their help.

Because the next Soyuz launch is delayed, that means Terry, Samantha, and Anton will be delayed in their return as well. They have each assured their space agencies that they are willing to stay on station as long as necessary, which I think reflects well on them, even if it's also true they have no choice. I know this must be stressful for them—we each know how long we'll be here and pace ourselves accordingly. I can't imagine having to call my family and tell them I'm not coming back when I'd said I would, and that I have no idea when I'll return. I can only sympathize with my crewmates. Outwardly, they all appear professional and upbeat. Terry tells me he sees this as a positive thing: it's a privilege to live in space, and now he gets to stay longer and complete more of the things he wanted to do, like taking pictures of specific places on Earth and filming an IMAX movie he had a particular fondness for. Samantha's attitude is more casual. "What are you going to do?" she asks, then points out that she will likely exceed the world record for the longest single spaceflight by a woman, 195 days.

When I finally finish the huge job of changing out the urine processor, it is satisfying knowing that we will be able to process urine and make clean water. But it's also strangely unsatisfying in that all I've done is put everything back the way it usually is. I reinstall the kabin, making sure all the tools are stowed properly, downlink the photos, then run on the treadmill for half an hour.

While I'm running, a smoke alarm annunciates loudly. The treadmill under my feet stops automatically. The emergency signals are designed to get our attention, and they do. Even as I'm unhooking my running harness and scrambling to respond to the alarm, I'm pretty sure I know what caused it—as I was running I probably liberated some dust from the treadmill or maybe caused the motor to smoke a bit by pushing against the treadmill in an effort to get my heart rate up. The fire alarm also automatically shuts down ventilation in Node 3, and that shuts down our Seedra. After we're fully recovered from the alarm, the ground informs us that they can't restart the Seedra and aren't sure why. I'm less than thrilled by the prospects of rising CO_2 until we can get it running again.

I've been looking forward to my videoconference with Amiko all day. Once a week we get to see as well as hear each other, for a length of time varying from forty-five minutes to an hour and a half. We have developed a ritual at the end of each videoconference: Amiko picks up her iPad and carries it around the house so I can see inside each room. It makes me feel connected to home to see our sofa, our bed, the pool, the kitchen—all of it flooded with sunlight, each of the objects held down by gravity. Once, in the kitchen, I noticed a warning light on the fridge—the water filter needed to be changed. I pointed it out to Amiko so she could have clean water too.

Today, I can see Amiko sitting on our sofa with light streaming in the window to her right. We talk about how each of our days has gone so far, then she reminds me about next week's videoconference: she has offered to have some of my friends over to our house so I can visit with them. She mentions that in the course of preparing for guests, she discovered that the speakers by our pool aren't working, and she hasn't been able to troubleshoot the problem on her own yet.

"I'll figure it out before Saturday," she says.

"Let's just fix it now," I suggest, and within minutes she's pointing the iPad camera at the web of cords on the back of the stereo components in the closet while I squint at the fuzzy image on my screen, trying to figure out which connection isn't working.

"Push that button on the left," I suggest. She tries to comply. "No, not that one, the one next to it."

"I'm pushing it," she says. "It's just not working."

The videoconference ends abruptly when we lose comm coverage. The image on my screen is static—the larger window showing Amiko, her face drained and expressionless in the dark closet; the smaller window showing my own face, frozen around the words I was saying. We both look incredibly annoyed. What if this were to be the last time we saw each other? I stare at the two faces for a moment, then turn off my laptop. The CO_2 is climbing, and I can feel the accompanying headache coming on.

A couple of hours later, when we have coverage again, I call Amiko's cell.

"I just wanted to tell you I'm so sorry I wasted our videoconference trying to fix the speakers," I say. "I should have just left it for later."

"I know you don't like to leave things undone," she says. The warmth is back in her voice. We talk for a bit longer, then say good night.

The next day I suggest to Amiko that she could download the stereo manuals from the manufacturer website, which made it easier to fix the problem. The following week, our videoconference party goes off without a hitch.

STILL NO WORD from the Russians on why the Progress malfunctioned. We don't know whether they have a good theory and just haven't confirmed it or if they actually have no clue whatsoever. Terry, Anton, and Samantha still don't know what their landing date will be. Every afternoon Terry floats over to the Russian segment through the dark, angled corridor of PMA-1 and into the FGB, over the tons of cargo strapped to the floor. Arriving in the open space of the service module, Terry pauses to look out the three Earth-facing windows in the floor that make the module feel like a glass-bottom boat, then asks Anton, who is always hanging halfway out of his crew quarters working on his laptop with his headset on, whether he has heard anything new about the Soyuz return. According to Terry, Anton shrugs and says no. Gennady tells us Moscow has identified a possible culprit for the Progress failure. He also tells us that our Soyuz, the one we came up here in and that Gennady will take back down with two other people in September, might have the same issue—a sobering thought that we could have been hopelessly lost. Not good news.

Since the ground was never able to get the Node 3 Seedra working again after the fire alarm, today Terry and I are working together to repair it. The experience is kind of like doing a transmission overhaul—a complicated, absorbing, detailed job—but our lives just happen to depend on this one. The other Seedra has not been operating smoothly, which puts a lot of pressure on us to make sure this one is repaired.

Dismantling the beast with Terry's help is much better than struggling with it on my own, but it's still unbelievable how much of a pain in the ass this thing is. The valves are positioned in places where no human hand can reach them, and we have to use four different-size wrenches, each turning a bolt only ten or twenty degrees, multiple times. It takes half an hour just to remove one bolt, and Terry tears up the back of his hand so much in the process he has to bandage it—in space, blood wells up into spheres and, if liberated, floats everywhere. We are finally able to get the Seedra out of its rack and float it to the Japanese module, where there is more room to work. Moving such a large mass is a slow and deliberate process. After a break to eat lunch, we return to finish the job. The next day, once we think we have it fixed, we return Seedra to Node 3 to reinstall it into its rack. It doesn't fit. We try different angles, different approaches, use more or less force, jiggle it, bang into it with our shoulders. Gennady comes down to add some extra muscle. Nothing works. Terry and I examine the beast and realize that there are some washers on the bottom of it that seem to serve no other purpose than to hold it in place once it's seated correctly. (They were probably designed to protect the Seedra from the vibration of launch.) If we remove those, I think it could shift down a bit and settle properly into place.

I call down to the ground to describe my idea about the washers, expecting to get the typical NASA answer that this will require further study and consultation with experts—days of emails, phone calls, and meetings—before they reach the conclusion that it would be an acceptable solution. NASA's tendency toward an abundance of caution and excessive analysis is both a good thing and a bad thing. We always err on the side of doing things the way they have always been done if those things haven't killed any astronauts or destroyed any important hardware. Yet this attitude often keeps us from trying new things that would save everyone a good deal of time and trouble. I don't think the control center always takes into account that our time and energy are resources that can be wasted.

After a short interlude of consultations, the ground tells us to try removing the washers. Terry and I exchange a surprised look. Maybe the culture in the control center is changing; maybe the flight controllers are getting better at trusting astronauts' judgment.

Having been given the go-ahead, I happily pry off the washers using a crowbar and a good deal of effort. Terry has to steady the Seedra in place while I pry, since in weightlessness the mass of the machine doesn't hold it down against the force I'm applying. Terry and I now can slide the Seedra into its rack perfectly, and the thunk it makes as it slides into position is deeply satisfying. We'll wait until tomorrow to try powering it up.

As we are putting our tools away, Terry shouts something with a childlike excitement in his voice: "Hey! Candy!"

A little piece of something edible looking is floating by. It often happens that bits of food get away from us and provide an unexpected snack for someone days later.

"Remember the mice," I warn him, "It might not be chocolate."

He takes a closer look at it. "Shit, it's a used Band-Aid," he says. He catches it and puts it in the trash. Later that night, we tell Samantha the story and she tells us that last week *she* ate something she thought was candy and realized only too late that it was garbage.

That night, floating in my sleeping bag with my eyes closed, I have one of those little convulsions people sometimes get when they are just about to fall asleep, when it feels like you're falling and you try to catch yourself. In space, these are more dramatic because without gravity holding me to the bed, my body undulates wildly back and forth. And this one was *especially* dramatic because it coincided with a bright cosmic ray flash. As I try to fall asleep again, I wonder whether the cosmic ray somehow triggered my reflex response or if it's strictly a coincidence.

In the daily planning conference, we learn that Terry, Samantha, and Anton will leave on June 11, more than a month late, and the new crew will come up on July 22. Their Soyuz has been docked here since November, and it's only safe for the spacecraft to sit idle for a certain period of time. It's not clear how much of this decision hinges on that time constraint, and how much on the determination that their Soyuz is free of the issues that doomed Progress. Either way, the Russian space agency has weighed the risks and decided it will soon be time for them to go.

After the daily planning conference, I immediately go through the steps of preparing the Seedra to be powered up. When I tell the ground we are ready, there is a dramatic pause.

"Powering up," capcom says. "Stand by."

We stand by.

It doesn't work.

"Son of a . . . bitch!" I say, being sure not to key the microphone since we're on an open channel.

"We'll take a look at this and get back to you," says the capcom.

"Copy," I reply, dejected.

Because it's Friday, we will have to live with high CO_2 levels all weekend; when one Seedra fails, it takes a while for the other to come up to speed, and flight controllers won't even start trying to figure out what's wrong until Monday. I'm going to feel like crap all weekend, and it will be even worse because that will be a constant reminder of what a clusterfuck this CO_2 situation is, how little the ISS program managers seem to care about our symptoms.

I knew this year was going to test my psychological endurance more than physical, and I think I was as prepared as anyone could be. Having flown a long-duration mission before, I understand how important it is to manage my energy from day to day and week to week, which includes choosing what to get upset about. But this is incredibly depressing. I float into my CQ to take a few minutes for myself and be pissed.

I click through some emails, aware that I'm using a bit more force on the laptop than is necessary. There's one from Amiko wishing me a happy Friday, and I decide to call her before heading over to the Russian segment for dinner. She picks up on the second ring and sounds happy to hear from me. She's still in the middle of her workday but is looking forward to the weekend. I try to keep the annoyance out of my voice, but she sees right through me.

"What's wrong? You don't sound good," she says. Even before I can draw breath to answer, she asks, "The CO_2 is high, isn't it?"

"Yeah," I say. I tell her the whole saga with the Seedra and what we face for the weekend. I tell her I'm impressed that she could tell the CO_2 was high just from my voice.

"Not just your voice, but your mood," she clarifies. "When you sound like you're letting things get to you, I know the CO_2 is high."

She is the only person on planet Earth who seems to care.

At Friday night dinner, we talk about the new landing date for Terry, Samantha, and Anton. I will be alone for six weeks on the U.S. segment before their replacements arrive. It's a long time to be floating around by myself, but being alone doesn't seem like a bad thing. I like having crewmates, and I've especially enjoyed working with Terry and Samantha, but being alone won't be an unwelcome change. Besides, each time people leave or arrive marks another milestone of my mission that I've successfully put behind me.

As we eat, I say, "I guess I'll be able to float around naked in the U.S. segment."

"You can float around naked now, if you want to," Samantha says with an offhanded shrug, digging in the bottom of a bag of ravioli.

"Guys, do you think the Soyuz landing will definitely be in June?" Anton asks Terry and me.

Terry and I look at each other, then at him.

"Anton, aren't you the Soyuz commander?" Terry asks rhetorically.

"*Da,*" says Anton. He shakes his head and smiles, acknowledging the strangeness of the situation. We should be asking him for information, not the other way around. "I thought you might have heard something I didn't." At times, it seems the Russian space agency deliberately keeps their cosmonauts in the dark.

"We'll let you know if we hear anything," Terry promises.

It seems as though we could use some better communication all around.

ALONG WITH Saturday morning science, we sometimes have other activities scheduled on weekends that weren't high enough priority to make it into the regular schedule. Today is one of those. Samantha is going to set up and test out a new piece of equipment designed by the European Space Agency: an espresso machine. Apparently when you have Europeans in space, you also have to have good coffee—the instant stuff just isn't the same. After working through the procedures to brew a small bag of espresso, including multiple troubleshooting calls to the payload operations center in Huntsville, the historic first espresso shot in space is brewed. I take a picture of Samantha holding the espresso in a special cup designed to allow sipping in zero g. As she takes the first drink, I say, "That's one small step for woman, one giant leap for coffee," over the space-to-ground channel. I'm pretty pleased with my line. The machine cost more than a million dollars to build, certify for flight, and launch; there are only ten espresso packets on board, making Samantha's drink a very expensive cup of coffee—worthy of a historic quote.

. . .

A USEFUL WAY to think of an orbiting object like the International Space Station is that it is going fast enough that the force of gravity keeps it curving around the Earth. We think of objects in orbit as being stable, staying at the same distance above the planet, but in reality the small amount of atmospheric drag that exists at 250 miles above the Earth's surface pulls on us even when we are whizzing along at 17,500 miles per hour. Without intervention our orbit would tighten until we eventually crashed into the Earth's surface. This will be allowed to happen some day when NASA and our international partners decide that the station has finished its useful life. It will be deorbited in a controlled manner to make sure that when it hits the planet, it will be in a safe area in the Pacific Ocean, and I hope to be there to watch. This is how the Russian space station *Mir* ended its life.

We keep ISS in orbit using a Progress that is docked here. Mission control calculates how long to fire its engine, and that force boosts us back into the proper orbit. Sometimes we wake up in the morning to learn that a successful reboost has taken place while we slept.

This morning, though, an attempted reboost failed. The Progress engine burned for just one second, not the several-minute burn we usually do. Once again, a Progress has failed to function properly, and, once again, we must worry about what that will mean for us.

We are not in any immediate danger of crashing into the Earth—it would take many months for our orbit to decay to a dangerous degree—but we also use the Progress engines to move the station out of the way of space junk, so the failure could have frightening consequences. This is another strike against a piece of hardware everyone had thought of as rock solid, challenging our confidence in the Soyuz spacecraft, which are made with identical or similar components and by the same manufacturer—including the one that is meant to be my ride home.

Now that we have lost the supplies that were supposed to reach us on the Progress, we have to be more vigilant about the trash we pack into the empty visiting vehicles, making sure we aren't disposing of anything usable. Terry and I spend some time going through bags of stuff that other crew members have discarded, looking for uneaten food, clean clothes, or other consumable supplies. While we work, we talk about whether or not Terry's Soyuz will leave anywhere near on time. As I'm sorting out food packets and talking, I find myself holding something made of fabric. It's some dude's used underwear. I stuff it into the trash and excuse myself to wash my hands a hundred times, an unsatisfying process without running water.

The good news is that the Node 3 Seedra is working again. It had failed because the fan that pushes air through the system wasn't starting. After some investigation and discussion, the ground devised a solution to fix it by replacing just the fan motor without pulling the whole unit out of the rack. That worked, miraculously, and now we are breathing clean air again. It's remarkable how good this is for morale.

That Friday night, we are having dinner on the Russian segment, and we know it will be one of our last with Terry, Anton, and Samantha. Terry floats to the U.S. segment to retrieve the last of the ice cream that came up on SpaceX, and when he comes back he has a troubled look on his face.

"Scott, the ground is trying to get in touch with you," he says. "You need to call your daughter Samantha right away. They said it's an emergency."

"Why didn't they call me here?" I ask. There is another space-to-ground channel in the Russian segment.

My crewmates all look at me with concern. They know that I got a similar call on the space station five years ago, when my sister-in-law was shot.

"I'm sure it's nothing," I say, for their benefit more than mine. I go to my CQ, where I can talk privately. Only then do I realize that we don't have communication coverage, and I won't be able to make a call for twenty minutes. I spend that time thinking about Samantha, about what she was like as a spirited toddler, as a bright-eyed school-age kid, as a moody teenager. I still blame myself for the problems Samantha and I have had in our relationship since her mother and I split up. The teenage and young adult years are a stormy time for a lot of kids, and I know that Samantha has had to deal with fallout from the divorce, caring for her mother and her younger sister in ways that I don't even know about. It's been an ongoing struggle to get to a place where we can be comfortable with each other without fear of blowups.

When the satellites are finally aligned, I put on my headset and click on the icon to place a call to Samantha's cell. She answers on the second ring.

"Hi, Dad." She knows it's me because calls from the space station are all routed through the Johnson Space Center.

"Are you okay? What's going on?" I ask, trying to sound calm.

"Not much," she says. "I'm at Uncle Mark and Gabby's. Everyone has left, and I'm lonely." I can tell from her tone that nothing is wrong. She sounds bored.

"That's it? There's no emergency?" I ask, feeling my concern subside and give way to irritation. It felt like the times I'd lost track of one of the girls at a shopping center and looked for them long enough to start fearing the worst.

Samantha explains that she had flown to Tucson for the high school graduation of her cousin Claire, Mark's younger daughter. Samantha had chosen to go to the graduation because she has been going through a hard time and was feeling cut off from our family while I am away. She thought it might make her feel better to be at a gathering of Kellys. But the night after the graduation Mark and Gabby had left town, and shortly after that, Claudia, Mark's older daughter, left as well, leaving Samantha by herself in an empty house. She had felt abandoned and wanted to get home, and when she didn't get a response to a number of emails, she had called Spanky. When he conveyed her request to mission control, her request had been misinterpreted as an emergency.

The absurdity is not lost on me that I'm in space for a year, and she's lonely. But I'm also reminded just how much my family is sacrificing for this mission.

She apologizes for scaring me and promises to leave a clearer message next time. I go back over to the Russian segment to rejoin the festivities, my mood somewhat dampened.

That night, I have one of those twilight falling-asleep dreams. For some reason, I'm focused on the death of Beau Biden, the vice president's son, who passed away from brain cancer yesterday at forty-six. I never met him, but I heard great things about him. His death bothers me more than I would have expected. In my half-awake state it occurs to me that one day we're all going to be dead, that we will all be dead much longer than we were alive. In a sense I feel I know what it will be like, because we were all "dead" once, before we were born. For each of us, there was a moment when we became self-aware, realized that we were alive, and the nothingness before that wasn't particularly objectionable. This thought, strange as it may be, is reassuring. I wake up long enough to type an email to Amiko about it.

People often ask me whether I had any epiphanies in space, whether seeing the Earth from space made me feel closer to God or more at one with the universe. Some astronauts have come back with a new view of humanity's role in the cosmos, which has inspired new spiritual beliefs or caused them to rededicate themselves to the faiths they grew up with. I would never question anyone else's experience, but this vantage point has never created any particular spiritual insight for me.

I am a scientifically minded person, curious to understand everything I can about the universe. We know there are trillions of stars, more than the number of grains of sand on planet Earth. Those stars make up less than 5 percent of the matter in the universe. The rest is dark matter and dark energy. The universe is so complex. Is it all an accident? I don't know.

I was raised Catholic, and as is the case in many families, my parents were more dedicated to their children's religious development than they were to their own. Mark and I attended catechism classes until one day in the ninth grade, when my mother got tired of driving us. She gave us the choice of whether to keep going or not, and, as many teenagers would, we chose to opt out. Since that day, organized religion has not been part of my life. When Samantha was ten years old, she asked me at dinner one evening what religion we were.

"Our religion is 'Be nice to other people and eat all your vegetables,'" I said. I was pleased with myself for describing my religious beliefs so concisely and that she was satisfied with it. I respect people of faith, including an aunt who is a nun, but I've never felt that faith myself.

. . .

WE WILL be spending a lot of time this week working on an experiment called "Fluid Shifts Before, During, and After Prolonged Space Flight and Their Association with Intracranial Pressure and Visual Impairment"—"Fluid Shifts" for short. Misha and I are the subjects of the experiment, and it promises some of the most important results for the future of spaceflight.

Maybe the most troubling negative effect of long-duration missions in space has been damage to astronauts' vision, including mine on my previous mission. At first, these changes were assumed to be temporary. Once astronauts started flying longer and longer missions, though, we showed more severe symptoms. For most, the changes gradually disappeared once the mission was over; for some, the symptoms seemed to be permanent. When I flew my first mission on the space shuttle, in 1999, I didn't need corrective lenses, but while on the mission I realized things were getting blurry in the middle range, ten or twelve feet—across the flight deck of the space shuttle. Back on Earth, my symptoms quickly resolved. My second flight was eight years later, by which time I had started using reading glasses. After about three days in space, I no longer needed them. The improvement lasted for about three months after I returned to Earth.

Three years later, for my first long-duration flight, 159 days, I was wearing bifocals all the time. After a short period in orbit, my vision got worse, and I wore stronger lenses to correct for the change. When I returned to Earth, within a few months my vision returned to what it had been when I left. But I had other troubling signs: swelling of the optic nerve and what seemed to be permanent choroidal folds. (The choroid is a blood-filled layer in the eyeball between the retina and the sclera—the white part—that provides oxygen and nourishment to the outer layers of the retina. These folds in the choroid could damage the retina and cause blind spots.) My vision symptoms so far this year seem to be similar to the last time, though we are monitoring them closely to see whether they will get worse.

If long-term spaceflight could do serious damage to astronauts' vision, this is one of the problems that must be solved before we can get to Mars. You can't have a crew attempting to land on a faraway planet—piloting the spacecraft, operating complex hardware, and exploring the surface—if they can't see well.

The leading hypothesis is that increased pressure in the cerebral fluid surrounding our brains is causing the vision changes. In space, we don't have gravity to pull blood, cerebral fluid, lymphatic fluid, mucus, water in our cells, and other fluids to the lower half of our bodies like we are used to. So the cerebral fluid does not drain properly and tends to increase the pressure in our heads. We adjust over the first few weeks in space and pee away a lot of the excess, but the full-head sensation never completely goes away. It feels a little like standing on your head twenty-four hours a day—mild pressure in your ears, congestion, round face, flushed skin. As with so many other aspects of human anatomy, the delicate structures of our heads evolved under Earth's gravity and don't always respond well to having it taken away.

The increased fluid pressure may squish our eyeballs out of shape and cause swelling in the blood vessels of our eyes and optic nerves. This is all still a theory, as it's hard to measure the pressure inside our skulls in space (the best way to measure intracranial pressure is a spinal tap, which I'd very much prefer not to have to undergo, or to perform on a crewmate, in space). It's possible, too, that high CO_2 is causing or contributing to changes in our vision, since it is known to dilate blood vessels. High sodium in our space diets could also be a factor, and NASA has been working to reduce that in order to test whether this makes a difference. Only male astronauts have suffered damage to their eyes while in space, so looking at the slight differences in the head and neck veins of male and female astronauts might also help scientists start to nail down the causes. If we can't, we just might have to send an all-women crew to Mars.

Since it's impossible to re-create the effects of zero gravity in a lab for sustained periods of time, scientists have conducted experiments on people with pressure sensors already installed in their skulls for other medical conditions. These people were taken up on an airplane that can create weightlessness for short periods in order to measure what happens inside their heads when they reach zero gravity. Their intracranial pressure dropped when they got to microgravity, rather than increasing as had been expected. Maybe it takes a while for the fluids to shift, or maybe the leading hypothesis is wrong. Before leaving for this mission, I volunteered to have a pressure sensor installed in my skull, but NASA declined my offer. The risks of drilling a hole in my head before sending me to space for a year were too great.

In the Fluid Shifts study, Misha and I will be subjects in an experiment that uses a device for relieving the intracranial pressure of spaceflight—pants that suck. This is not a metaphor. We will take turns donning a device, roughly the shape of a pair of pants, called Chibis (Russian for "lapwing," a type of bird), that reduces the pressure on the lower half of our bodies. The pants look a lot like the bottom half of the robot from *Lost in Space*, or like Wallace and Gromit's "wrong trousers." Reducing the pressure on our lower bodies also reduces the amount of fluid in our heads. By studying the effects of Chibis on our bodies, we hope to understand more about this problem.

One of the times these pants were used, however, the Russian cosmonaut wearing them experienced a sudden drop in heart rate and lost consciousness. His crewmates thought he was in cardiac arrest and immediately ended the experiment without ill effect. Anytime a piece of equipment has put a person at risk, NASA has been reluctant to use it again. But because the Chibis is still the best possibility we have for understanding this problem, they are making an exception.

Preparing to don the pants is actually a days-long process. We have to take baseline samples of blood, saliva, and urine, and we also have to take images of blood vessels in our heads, necks, and eyes using ultrasound. So much of the equipment we need to do these tests is only on the U.S. segment, so we spend a few hours packing it up and ferrying it over to the Russian service module. This is going to be the most complicated human experiment that's ever been done on the International Space Station.

When it's time to put on the device, I take off my pants and clamber into the Chibis pants, making sure the seal around my waist is secure. Misha is working the controls, slowly decreasing the pressure on my lower body, and with each incremental change I can feel the blood being pulled out of my head—in a good way. For the first time in months, I don't feel like I'm standing on my head.

But then the feeling starts to change. It's like I'm in an F-14 again, pulling too many g's. I can feel myself starting to gray out, my peripheral vision closing in, where you are at risk of losing consciousness. The pants are malfunctioning, and I feel like I could have my intestines pulled out in the most unpleasant way possible.

"Hey, something's not right with this," I announce to Misha and Gennady. "I'm gonna have to—" I reach for the seal at my waist, prepared to break it, canceling the experiment. At the same instant, I hear Gennady yelling.

"Misha, *shto ty delayesh*?" What are you doing? Gennady doesn't yell much, so when he raises his voice you can be sure you have likely screwed up. In this case, I look over at the pressure gauge, which is not supposed to go past 55. Misha has it down to 80, the maximum negative pressure.

Fortunately neither I nor the equipment sustains any permanent damage, and we are able to go on with the experiment. I stay in the pants for a couple of hours, doing various medical tests like measuring blood pressure and taking ultrasound images of my heart, neck, eyeball, and a blood vessel just behind my temple. This is where my space tattoos come in handy. Shortly before my launch, I visited a Houston tattoo parlor and had some black dots placed on the most-used ultrasound sites (on my neck, biceps, thigh, and calf) so I wouldn't have to locate the exact spot each time. It's saved me a huge amount of trouble already. We measure my cochlear fluid pressure (by sticking an instrument in my ear) and my intraocular pressure (by tapping a pressure sensor on my anesthetized eyeball). We scan my eyeball with a laser, which can register changes like choroidal folds and optic nerve swelling.

During the time we're doing this, I feel as good as I've felt in space. The constant pressure in my head clears, and I'm sorry when it's time to shed the pants and shut the experiment down.

Later in the day, I'm sitting in the Waste and Hygiene Compartment. I've been sitting for a while, in fact—sometimes this process takes a while in the absence of gravity. Samantha is brushing her teeth just outside the kabin, which is like a stall in a public restroom—and I can hear her humming to herself, as she often does while she works. I can see her socked feet under the wall, hooked under a handrail to keep her steady. Her toes are close enough that I could reach out and tickle them, but I decide against it.

This scene probably sounds a bit odd to those who haven't experienced the loss of privacy on a space station, but we get used to it. I've just been reading about how the men on the Shackleton expedition had to hunker down behind snow drifts and had only chunks of ice to clean themselves with, so I count myself lucky. Because I have nothing else to do while I sit, I watch Samantha's feet hooked under the handrail, keeping her body perfectly still, and I think about the complexity of that simple task. If you showed me nothing but a foot hooked under a handrail in zero gravity, I could estimate how long that person has been in space with a high degree of accuracy. When Samantha was new up here, she would have hooked her feet too hard, used too much force, and tired out her ankles and big toe joints unnecessarily. Now she knows exactly how little pressure she needs to apply. Her toes move with the elegance and precision of a pianist's fingers on a keyboard.

Last night we enjoyed our final Friday dinner with Terry, Samantha, and Anton. Since the loss of the Progress, the Russians are running low on food and other supplies, and though we've made it clear we will share food, things won't be the same for a while. I bring over a salami my brother sent up on the last SpaceX, and I eat some of the last of the Russian meals, a "Can of White" (chicken with white sauce), and some American "Bags of Brown" (some sort of irradiated beef thing). The Russians also have something called "the Appetizing Appetizer," which it is not.

A few of us say we have been craving fruit recently, which is no surprise given that there has been no fresh food in our diets since shortly after Dragon arrived. Our dried, bagged, and canned fruits are not the same as the real deal. I share the fact that I recently had a craving for a cheap domestic beer in a small bar glass with warm, bitter foam like my dad used to drink. This craving is weird, because I haven't had that kind of beer since college and would never choose to drink it on Earth. I'm more of a hoppy India pale ale kind of guy. Maybe there's some nutrient in cheap beer I'm missing. We talk about whether we are going to get scurvy, and what it is exactly, what the symptoms are. I scratch my balls to get a laugh. Just the word "scurvy" sounds horrible, we agree. I wonder whether the members of the Shackleton expedition got scurvy; I will look at the book again tonight before I go to sleep. When the next SpaceX resupply gets here at the end of June, it will bring fresh fruits and vegetables as well as desperately needed supplies, chief among them the shit cans that are so vital to life in space. My brother has also announced he is sending me a gorilla suit on SpaceX. I asked why I needed a gorilla suit on the space station.

"Of course you need a gorilla suit," he responded. "There's never been a gorilla suit in space before. You're getting a gorilla suit. There's no stopping me."

I'm concerned about devoting cargo space to something that seems frivolous. There are those who look for reasons to criticize NASA and any expense that appears to be excessive, and I know those people would get out their calculators to figure the cost for sending a gorilla suit into orbit. Mark tells me that after being vacuum-packed for flight in space, the gorilla suit is no bigger or heavier than a sweatshirt we might send up as a shout-out to an alma mater or an organization.

As we finish dinner, we talk about all we have accomplished on this expedition: the visiting vehicles (including the ones that didn't make it), difficult and risky maintenance on the spacesuits, important life-science experiments, and the rodent research, which we will finish up the day after tomorrow. We also talk about our evolving relationships with the various control centers—Houston, Moscow, Europe, Japan—and how much the mutual adoration society, as I call it, has gotten out of control. It seems that no one can do anything, either in space or on the ground, without receiving a short speech of appreciation: "Thank you for all your hard work and your time on this, awesome job, we appreciate it." Then the speech has to be repeated back: "No, thank *you*, *you* guys have been just awesome, we appreciate all *your* hard work," ad nauseam. It all comes from a well-meaning place, but I think it's a waste of time. I've often had the experience of finishing up some task and then moving on to the next thing, when a "thank you" speech comes back at me. This requires that I stop what I'm doing to float back to the mic, acknowledge those thanks, and return them in roughly equal proportions—multiple times a day. If you consider the cost of constructing and maintaining the space station, the mutual adoration society probably costs taxpayers millions of dollars a year. I'm already thinking about putting a stop to it when Terry, Samantha, and Anton leave.

On Wednesday, the day before the Soyuz is to leave, Terry must hand over command of the station to Gennady. There's a little ceremony, a military tradition drawn from the Navy change-of-command ceremony, that lets everyone know clearly when responsibility for the station transfers from one person to another. The six of us float somewhat awkwardly in the U.S. lab while Terry makes a speech. He thanks the ground teams in Houston, Moscow, Japan, Europe, and Canada, as well as the science support teams in Huntsville and other places. He thanks our families for supporting us on our missions.

"I'd like to say a few words about the crew I launched with," Terry says, "Anton and Samantha, my brother and sister." This might sound a bit exaggerated, but I know from experience how flying in space as a crew brings people together. Terry would do anything for them, and they for him. "We got to spend two hundred days in space together, including a few bonus days, and I couldn't have asked for a better crew.

"So now Expedition Forty-three is in the history books, and we turn it over to a new chapter and Expedition Forty-four." With that, he hands the microphone to Gennady, who checks to see if it is still on.

"No matter how many flights you have," Gennady says, "it's always like a new station, always like first flight."

This makes everyone smile, because Gennady has more spaceflights than any of us (this is his fifth), and he will soon set a record for most days in space of any human. Gennady wishes Terry, Anton, and Samantha a "soft, safe landing and the best return home." Terry tells the control center that this concludes the handover ceremony, and another milestone of my mission is crossed off. The next handover ceremony will be in September when Gennady leaves and I become commander.

Later that night, Terry asks me what landing is like in the Soyuz. He's trained for this, of course, and he has been told what to expect by Anton and by the training team at Star City; still, he is curious to hear my experience. I think of how to set him up for what to expect without scaring him too much.

We call Samantha over so she can hear it too, and I describe what my experience had been last time: As we slammed into the atmosphere, the capsule was engulfed in a bright orange plasma, which is a little disconcerting, sort of like having your face a few inches away from a window while on the other side someone is trying to get at you with a blowtorch. Then, when the parachute deployed, the capsule spun and twisted and turned violently in every direction. If you can get in the right frame of mind, if you can experience it like an adventure ride, this can be great fun. On the other hand, some astronauts and cosmonauts, after their first Soyuz landing, have said that they were being thrown around so violently they became convinced something had gone wrong and they were going to die. There can be a fine line between terror and fun, and I want to give Terry and Samantha the right mind-set.

Terry has experienced the ride back to Earth on the space shuttle, and I tell him the Soyuz reentry is much steeper. "The shuttle reentry feels like cruising down Park Avenue in a Rolls-Royce," I tell him. "Riding the Soyuz is more like riding a Soviet beater car down an unpaved street that leads off a cliff."

They both think this analogy is funny, but they also appear a little worried.

"As soon as you realize you aren't going to die, it's the most fun you'll ever have," I tell them. "I'll tell you the truth—the ride is so much fun, I would sign up for another long-duration mission just to get to take that ride again." Terry and Samantha look skeptical, but it's true.

. . .

OUR CREWMATES ARE leaving today. There is a ceremony for the hatch closing, seen live on NASA TV, as they depart. It starts out a bit awkwardly, since all six of us are crammed into the narrow Russian module where their Soyuz is docked. I snap some pictures of Anton, Samantha, and Terry posing in the open hatch. Then those who are staying wish them good luck and a soft landing. Anton hugs Gennady, whom he looks up to so much. Then he hugs Misha. Then he hugs me. Samantha hugs Gennady, then Misha, then me. It seems to me that Samantha gives me an extra-big hug, and after she has disappeared I realize that I won't be in the physical presence of a woman again for nine months. The three of them float into the Soyuz and give one last wave while we take their pictures.

Anton and Gennady wipe down the hatch seal in the vestibule, to make sure that no foreign objects keep the hatch from sealing properly. Gennady closes the hatch on our side while Anton is closing it from their side. And that's it. It reminds me of seeing off Charlotte at the airport at the end of a visit—after spending so much time together, I give her a hug, watch her walk down the jetway, and after a final wave, she disappears. It's a weird thing: I've spent so much time with these people, but with a few good-byes and hugs, our shared experience is over in an instant.

I'm not scared for my departing crewmates, any more than I'm scared for myself, but seeing the hatch close behind them gives me a strange sense of isolation, even abandonment. If I have to work on the Seedra again, I'll have to do it without Terry's help. If I get into a discussion with the Russians about literature, I'll have to do it without Samantha's help. I'm looking forward to having the U.S. segment to myself, though, and I try to focus on that.

I float off toward the U.S. lab, and the Russians float off to their segment, and then all is silent. It's just me and the fan noise. No talk from Terry, whose upbeat commentary has punctuated everything I've done since I've been up here. No quiet humming from Samantha. For the moment, I don't even hear any voices from the ground.

I look around the junk on the walls in the U.S. lab, which suddenly feels much larger. I have the strange feeling I meant to say something more to Terry or Samantha, that I wanted to remind them about something, but I can't think what.

Then I hear Terry's voice, breaking in midsentence, as if he were here with me: ". . . pills for the fluid loading protocol, Anton? Or did you leave them on station?"

"I've got them," Anton answers, then rattles off a series of numbers in rapid-fire Russian to their control center. Now that the communications on the Soyuz are set up, I can hear through our intercom system every word my former crewmates say as if I were in there with them. I join the space-to-ground channel to warn Terry that his mic is hot and that everyone with an internet connection or tuned to NASA TV can hear every word he says. I wouldn't want one of them to inadvertently drop an F-bomb and then have to hear about it when he or she gets back to Earth. (Since inadvertently dropping the F-bomb to Earth myself, I am sensitive to the nuances of our comm system. On my second shuttle flight, I said "Fuck" while struggling with a piece of hardware in the airlock. My crewmate Tracy Caldwell called out, "Hot mic!" from the flight deck to let me know I could be heard on NASA TV. "Shit!" I said in response, making two FCC violations in ten seconds.)

I go through the rest of the afternoon listening to Terry, Anton, and Samantha's voices. As I work on a physics experiment, I can hear Samantha humming absentmindedly. A couple of times I turn around to say something to her, then remember where she is.

When the Soyuz is ready to detach and push away from station, three hours after we closed the hatch, I watch its departure on a laptop screen on NASA TV, just as many people on Earth are doing. I grab a mic.

"Fair winds and following seas, guys," I say. "It was a real pleasure spending time up here with you, and good luck on your landing."

Terry answers, "Thanks, Scott, we miss you guys already."

Gennady adds from the Russian segment, "Samantha, I think you forgot your sweater."

I hear them talking to one another this way, trading idle work chat and calling out numbers to the control center, almost all the way to the ground. If I didn't know what they were doing—falling like a meteor at supersonic speed toward the planet's surface—I could never have guessed.

Several hours later, they are on the ground safely in Kazakhstan. They had been here with me twenty-four hours a day for months, and now they are as far and unreachable as everyone else on Earth, as Amiko and my daughters and the 7 billion other humans.

That night, when I turn out the lights and climb into my sleeping bag, I'm aware of the quiet. There is no rustling in the other crew quarters or quiet talking as crewmates communicate with the ground or say good night to their families on the phone. If this were a normal six-month flight I would already be halfway done, but instead I feel I have as long as I did when I first got up here. Nine months. I don't often let these kinds of thoughts into my head, but when they do it's hard to get them out again. What have I gotten myself into?

SUNDAY RARELY FEELS like a Sunday on the space station, but today might be an exception. Yesterday I did both my weekly cleaning and my exercise, so today I actually have the entire day off. When I wake, I read the daily summary that was sent to us overnight and see that today Gennady sets the world record for the most days in space: 803. By the time he leaves, he will have 879, a record I expect to stand for a long time. I sleep late, eat breakfast, read a bit, then decide to clean out my email inbox. But when I open my laptop, there is no internet connection. This has been an ongoing problem: on Saturday nights the ground reboots the laptops remotely, and no one notices that the internet connection has been dropped. When I call down to ask for it to be fixed on Sunday morning, I'm told that the only person who knows how to do it doesn't come in until later in the day.

There is a SpaceX launch scheduled today for 2:20 p.m. our time (10:20 a.m. in Florida), and I had looked forward to watching it live, but my internet connection won't be fixed by then. SpaceX is carrying a lot of things we are looking forward to getting, most important being an International Docking Adapter, a $100 million mechanism that will convert docking ports built for the space shuttle to a new international docking standard, agreed to in 2010 by NASA, ESA, Roscosmos, the Japanese space agency, and the Canadians. (Ultimately it could even be used by China or other nations.) Without these adapters in place, we wouldn't be able to bring people up on SpaceX or the Boeing spacecraft still under development.

Also on board SpaceX: food (the Russians are still running low); water; clothing for American astronaut Kjell (pronounced "Chell") Lindgren and Japanese astronaut Kimiya Yui, who will both arrive next month; spacewalk equipment for Kjell, who will be my spacewalking partner in the fall; filtration beds for removing contaminants from our water (which is close to undrinkable with increasing levels of organic compounds, since the last set of beds, which we badly needed, blew up on Orbital); experiments designed by schoolchildren (some of the kids who saw their experiments blow up on Orbital are being given a second chance to see their work go to space today).

Personally, I'm looking forward to an extra set of running shoes, another harness for the treadmill, clean clothes, medications, and crew care packages that my friends and family chose for me.

Launch time comes and goes. Shortly after, my laptop's internet starts working again. I look up the video for the SpaceX launch, but the connection isn't strong enough to stream the video. I get a jerky, frozen image. Then my eye stops on a headline: "SpaceX Rocket Explodes During Cargo Launch to Space Station."

You've got to be fucking kidding me.

The flight director gets on a privatized space-to-ground channel and tells us the rocket has been lost.

"Station copies," I say.

I take a moment to think over all the stuff that has been lost. Kimiya's underwear, my pills, NASA's $100 million adapter. Schoolchildren's science experiments. All blown to bits. I joke to Mark that the thing I'm saddest about is the gorilla suit. After having to be talked into it, I had started thinking about all the fun Space Gorilla could have up here. Now he is a burned cinder and raining into the Atlantic Ocean, like everything else on the spacecraft. As stunned as I am by the loss, as overwhelmed as I am by what this will mean for the rest of my year in space and beyond, I'm almost as annoyed that I didn't get to watch the launch—and the explosion—live. I feel oddly left out of something that is having a huge impact on my life.

I call Amiko and she fills me in on what it looked like: two minutes after launch, the rocket reached maximum aerodynamic pressure, as it was supposed to, then it suddenly blew up in the clear Florida sky. As we talk, it starts to sink in that we have lost three resupply vehicles in the last nine months, the last two in a row. Our consumables are now down to about three months' worth, and the Russians are much worse off than that.

It occurs to me that maybe we should delay the next crew's launch until after the increment in September when, for a brief period, we will have nine people up here, with limited supplies and sky-high CO_2. It also occurs to me that the ground should have listened to me when I suggested Terry leave his spacesuit gloves for Gennady to use if we have to do an emergency spacewalk. New gloves are coming up on SpaceX, I was told dismissively. Now those gloves are flaming bits off the coast of Florida.

I think about the schoolchildren who saw their experiments blow up on Orbital, rebuilt them, and saw them blow up on SpaceX. I hope they will get a third chance. There is a lesson here, I guess, about risk and resilience, about endurance and trying again.

8

In the spring of 1988, I moved to Beeville, Texas, a small dusty town of blowing tumbleweeds halfway between Corpus Christi and San Antonio. Beeville is one of a few centers of the universe for young Navy pilots who want to fly jets, and I was thrilled to be there. I moved into a small ranch-style house on a dirt road across the street from a cattle ranch with two college classmates who were also in flight school, ready to start my training.

I began flying the T-2 Buckeye, a twin-engine jet. The first time I dressed in a G suit and oxygen mask to climb into the cockpit, I felt like I had arrived in the big leagues. The T-2 is a forgiving jet, which is why we trained on it first, but it's a jet just the same, which is to say challenging and dangerous to fly. I had a lot to learn. A jet has a lot more power than a propeller-driven airplane. It can go faster, it can accelerate quicker, and it is more responsive to the pilot's touch—all of which make it much easier to "get behind" the airplane (when it feels like the airplane is in control rather than the pilot) and get into trouble.

I had to get used to the feeling of wearing an oxygen mask and G suit and flying while strapped into an ejection seat. The equipment is physically restrictive, and wearing it made me more aware of potential danger. It was more intimidating than I had anticipated. At the same time, in that G suit I tended to hold my head higher, shoulders back, and walk with a spring in my step. I was becoming a tactical jet aviator and I was proud of it. There would be times in the near future, though, when my cockiness would be dealt a blow.

After I had flown that airplane for about a hundred hours, it was time to try landing on an aircraft carrier—a Navy ship with a flight deck to launch and recover airplanes. Because an aircraft carrier's flight deck is so short, it is equipped with catapults to help the aircraft take off and arresting cables to help them stop. The landings are difficult and dangerous, even under the best of circumstances.

This is the point in training when a lot of pilots wash out. I'd known this from the start, thanks to *The Right Stuff.* Carrier qualifications would be flown out of Pensacola, so I flew there the day before and met my brother and some of his squadron mates at McGuire's, the bar with the dollar bills all over the place. Mark was a year ahead of me, since I had repeated my freshman year of college. He had gone to Corpus Christi for flight training and was now finishing up qualifying to land the A-6 Intruder on the carrier. When I met him and his squadron mates at the bar, they were all celebrating because Mark and a few other guys had just qualified for both day and night landings on the ship. Now that he had qualified, Mark would soon be moving on to his fleet squadron, stationed in Japan.

The flight deck of an aircraft carrier is an incredibly dangerous place. It's not uncommon for people to be killed or seriously injured there, despite the high level of training. People have died walking into spinning propeller blades, getting sucked down a jet intake, or blown over the side by a plane's exhaust. Much of the operation is done by a bunch of teenagers, and to avoid accidents everyone must know exactly what his or her job is and perform it well. Mine was to land the plane.

The weather was not great, and because of my experience level I wasn't yet allowed to fly in cloudy conditions. As I got closer to the ship while keeping an eye on the weather, I noticed my roommate in another T-2 close by. I told him that to avoid running into each other as we dodged the clouds, I would join up on his right wing and we would fly formation. This was against the rules—neither of us had sufficient experience flying formation—but it seemed like the safest thing to do. Once we were clear of the bad weather, I backed off and fell in behind him as we approached the ship.

Looking down at the USS *Lexington* in the water, I couldn't believe I was going to have to land my jet on that tiny dot. When you land an airplane at an airport, the runway is generally at least 7,000 feet long and 150 feet wide. More important, though, it holds still. The runway on an aircraft carrier is less than 1,000 feet in length and much narrower—and it also pitches, yaws, rolls, and heaves along with the ocean's swells. The ship is also moving forward in the water, and because the landing area is angled with respect to the ship's bow, it is constantly moving away from and to the right of the jet trying to land on it.

The sight of the ship was intimidating. As I flew overhead and turned downwind, I didn't pull back on the stick hard enough, which drove me wide. This made it much more difficult to get lined up properly behind the ship. As I approached for my first landing, the deck actually seemed big compared to my T-2, which was deceiving, because my landings still weren't very precise. I tried to look at the optical landing system at the left side of the flight deck, a visual aid that lets pilots know how accurate their approach is. I hit the deck and added full power, heading back off into the air. My first attempt hadn't gone badly, and now I was slightly more confident. I was to do six touch-and-goes—landing and taking off again immediately—before extending the plane's tailhook to grab the arresting cable on the flight deck. I'd have to make four actual landings in order to qualify, and I hoped to make them all that day. As soon as I made my first arrested landing, I would officially be a carrier aviator, or "tailhooker," part of a unique fraternity.

I got through all my touch-and-goes with no problem. But when I put the hook down while approaching the ship, the danger of the situation became more real to me and I felt my adrenaline rising—not a good thing. I approached, touched down, and went to full power as we had been trained to do, in case the hook missed the wires—I'd need to be ready to leap back up into the sky to prevent my airplane from sliding off the front end of the aircraft carrier into the water. The feeling when the hook caught the wires and confirmed that I had done everything right would have been fantastic—if I hadn't forgotten to lock my harness properly. As my airplane was caught on the arresting cable, I was thrown forward and smashed into the instrument panel. The effect of getting into what felt like a car crash at the same time I had made my first heart-stopping carrier landing combined to slow down my reflexes. I was now supposed to reduce the power once I came to a stop, but I was having trouble doing it quickly. One of the aircraft handlers ran out in front of the jet, wildly giving me the "power back" signal.

I did a second arrested landing, then another one. One more and I would have the required four. But then it started getting dark, and we were sent back to the airfield. I expected to go out the next day and do the last required landing for my qualification, but when I saw I wasn't scheduled to fly, I had to assume I had disqualified. I was upset for a few hours, thinking that I had failed. But it wasn't long until I learned that I had done well enough on the three landings I completed that I was qualified without the fourth. I was a tailhooker.

SOON, I started flying the A-4 Skyhawk, an attack jet from the Vietnam era that let us learn more of the capabilities we would need for flying in combat: dropping bombs, flying at low altitude in order to evade detection, and air combat maneuvering. Just as in the T-34 and T-2, the pace of the training was aggressive. We were expected to learn quickly and move on to the next challenge. At around this point in the training, the pilots who had previous flight experience started to lose their advantage as the rest of us caught up. To learn how to drop bombs, we flew from Beeville to the Naval Air Facility El Centro in Southern California, two hours from San Diego, which is set up with targets for pilots to practice on. I wasn't especially gifted at dropping bombs, and nothing I tried to improve my accuracy seemed to work. I got used to taking ribbing from my classmates about it, but I wasn't the worst; occasionally someone else would drop a practice bomb so far off target it would get close to the spotter sitting in a shack at the edge of the bombing range.

The targets had been given strange names, probably so we could differentiate them from one another on the radio. Some of them I still remember: Shade Tree, Loom Lobby, Inkey Barley, Kitty Baggage. The targets were set up with different run-in lines to let us practice different approaches over different terrain. Each target consisted of concentric rings with a clearly marked center point that we would try to hit with our Mark 76 practice bombs. The A-4 bombsight was a fixed reticle, a light projected onto the windscreen, and using it required that I not only hold that dot on the target but also visually compensate for wind. I released the bomb by pushing a small button on the stick, and I had to account for the time it took the bomb to fall from my altitude. The temptation was to fly lower, decreasing the variables of the fall, but I couldn't drop so low I would be in danger of crashing.

I took much more naturally to air combat maneuvering, otherwise known as dogfighting. We started off with the basics, flying behind the instructor's aircraft in a position to be able to fire the gun, then trying to stay there as the instructor's airplane started moving around unpredictably. This was humbling at first, as the instructor was somehow able to go from being in the defensive position (in front of me) to an offensive position (behind me). I quickly got the hang of it, though, and as the engagements got more complicated, I gained in confidence. Thinking in three dimensions, as you must when dogfighting, came naturally to me. I soon learned the validity of the naval aviator motto: "If you're not cheating, you're not trying hard enough." I learned that if I showed up at the point where our engagement started with more air speed than I was supposed to have, I had a slight advantage.

This was one of my favorite phases of training, not only because I did well at it, but because it was fun. I experienced a freedom and creativity in air-to-air "combat" that I hadn't found anywhere else. I loved getting into long rollers, maneuvering the aircraft up and down amidst the large billowing cumulus clouds of the early Texas summertime, trying to "kill" my opponent. The last flight in this phase of my training I gave one of the instructors an epic beat-down—at least that's how it felt to me.

After I successfully carrier qualified in the A-4, I got my aircraft assignment. I was to fly the greatest Navy fighter plane ever, the F-14 Tomcat.

I had been in Beeville for about a year when I got my wings. My parents came for the occasion (my brother was unable to attend because of his own Navy duties). We lined up in our white dress uniforms for the ceremonial pinning on of our wings. My mother pinned my wings on me, a glowing, proud expression on her face. I remembered the day she had graduated from the police academy, when I got to see her lined up with her classmates in uniform, and the impression that sight had made on me. Now things had come full circle.

I WAS ASSIGNED to Fighter Squadron 101, the Grim Reapers, and moved to Naval Air Station Oceana in Virginia Beach, Virginia, for initial training on the F-14 Tomcat. My roommate and I drove overnight and I started my training almost immediately. Just as I had done with other aircraft, I progressed quickly from familiarization training to formation training to basic intercepts—finding another airplane and locking onto it with the radar. Eventually I began to learn basic air combat maneuvers, and this was when we began to feel like true fighter pilots. I practiced flying against a similar airplane (another F-14), a dissimilar airplane (like the A-4 or F-16, the Navy's best approximation of the Soviets' MiG), and flying against different numbers of enemy planes. All of this training would culminate in taking the airplane to the ship, which would be much harder than it had been in the T-2 and the A-4, since the Tomcat had such poor flying qualities and we had to qualify at night.

There is no training version of the F-14; there is no stick in the backseat, meaning the instructor can't take over for the student. We did a lot of classroom work first, learning the systems of the airplane, then putting in many hours in the simulator before climbing into the cockpit for the first time.

My first two flights were with an experienced pilot in the backseat, who was memorable for the chew he had in his mouth at all times, including while flying the jet. He must have just swallowed the spit. After that, I flew only with an instructor RIO (radio intercept officer, like Goose in *Top Gun*). I found it odd to have someone who wasn't a pilot grading me on my flying skills.

We quickly advanced to learning to fly the airplane in combat: air-to-air gunnery, basic intercepts day and night, single- and multiplane air-to-air engagements and low-altitude flight training. Air-to-air gunnery involved a lot of airplanes flying in a pattern around another airplane, which towed a banner the others were trying to shoot at. These exercises were done using real bullets, which seems like a terrible idea, though I never saw anyone get shot by accident. Each of us had bullets painted with a different color so the instructors would be able to tell who had hit the target how many times. Just as with bombing, I wasn't particularly good at this, but I enjoyed the competitive aspect.

The night before I tried to land the F-14 for the first time on the USS *Enterprise* off the coast of Virginia, I lay awake in bed for a long time. Our instructor had told us, "You won't be able to sleep, so just try to lie still and think about nothing so you get some rest." This turned out to be good advice and has served me well many times in the years since.

My first arrested landing was a complete disaster. I landed so low that my tailhook hit the back of the ship. That's called a hook slap, and it's not good. Basically, if I'd been any lower I would have crashed, and that would have been the end of me and my RIO. While none of my subsequent approaches were as bad as that hook slap, I didn't get much better. After a while, the instructors had seen enough and sent me home. I had disqualified.

I landed back at Naval Air Station Oceana with a weird feeling of disbelief. After my RIO and I jumped out of the plane, he looked at me with concern. I must have appeared as bewildered as I felt.

"Hey, you'll figure it out," he told me with an awkward pat on the shoulder. "Don't worry about it. Shake it off."

I could only mutter in response. There had been so much riding on this, and I had failed. As I went inside and took off my gear piece by piece—helmet, harness, G suit—I couldn't believe how much I sucked at this. I didn't know how I could improve if I got another chance, and it was possible I wouldn't get one.

I thought about what I might do with my life if I couldn't fly jets in the Navy. I had once picked up an application to the CIA at a college fair. That might be interesting. I thought about the FBI—that is, assuming the Navy would discharge me rather than send me out to fly a heavy airplane, work on a ship, or, worst of all, fly a desk. I had a couple of weeks to think about what my alternatives were while the Navy deliberated over my fate.

In the end, they decided to give me a second chance. I started all over in the carrier qualification phase, where I was paired up with a RIO who had been given the call sign "Scrote" because some unkind squadron mates had decided his face looked like a scrotum. Scrote had a good reputation for helping pilots who were having trouble behind the boat like me.

"You know, you can fly the airplane okay, but you're not *flying* it all the time," he told me. "You're on altitude and airspeed, but you're not *on top* of it." I had been trained to keep my altitude within a two-hundred-foot range, so I didn't worry if I was ten feet off the precise altitude, or twenty, or fifty. But Scrote pointed out that this imprecision in the end would lead me far from where I needed to wind up, and fixing it would take a lot of my attention. I had to always be making small, constant corrections if I wanted to make the situation better. He was right. My flying got better, and I've been able to apply what I learned from him to a lot of other areas of life as well.

My second attempt to qualify was on a black night with no moonlight. As I got within a couple of miles of the ship, I felt the pressure of what I was about to do. I started peeking out from my scan of the instruments inside the cockpit to see whether I could spot the ship. It was disorienting to see the faint lights of the carrier in an ocean of black. At three-quarters of a mile, the air traffic controller told me to "call the ball," to start flying the approach visually (rather than using the aircraft's instruments). My first thought was, *Oh, shit,* but I flew as I had been trained to, made small corrections to power and lineup. The glow of the flight deck that had looked so dim from the air became brighter until it was an all-encompassing yellow haze, and the next thing I knew I felt the tug of the arresting cable. I felt I had arrived on some alien planet, a new landscape that looked absolutely surreal. I had landed safely and successfully on that dark night.

The day you qualify to land on an aircraft carrier is a big deal, and when you do it at night, it's an even bigger one. As with many things that were a big deal in my squadron, it was traditional to have a party to celebrate it. This party was at my house, a three-story condo a few blocks from the beach that I shared with two other guys. To prepare for the party, we bought a ton of beer, some chips, and some Jell-O for making Jell-O shots.

My roommate's girlfriend had brought a friend to the party—Leslie Yandell. I remember seeing Leslie sitting on my couch talking with her friends and drinking a beer. She was cute, with a bright smile and curly blond hair. I decided to talk to her for a bit and found out that she had grown up in Georgia but lived nearby. Her stepfather was a dentist, and she worked as a receptionist in his dental practice. She was easy to talk with, and I liked her laugh, so I asked her out for the next weekend. She said yes.

In the Navy at that time, there was an idea that a single officer generally didn't advance as quickly as a married one. This wasn't a written rule—maybe it was even nonsense, but everyone believed it. Being a family man was supposedly an indicator of a certain kind of stability and maturity. I knew that all of the original Mercury astronauts Tom Wolfe wrote about were married, with at least two children. I wanted a family, and now that I was twenty-six and advancing in my career, it was starting to seem like the time was right. My brother was already married, and that added to the feeling that I should be ready for this stage of my life.

Leslie and I dated regularly for most of that year. I enjoyed going to her mother and stepfather's for dinner every Sunday. (He had been a commander in the Navy Reserve.) I soon became close with Leslie's brother and sister-in-law as well. It seemed like a logical next step to officially become part of their family. I asked Leslie to marry me over a bottle of wine sitting on a park bench on the edge of the Chesapeake Bay, and she said yes.

In September 1990, I was assigned to a real fighter squadron, VFA-143, nicknamed The World Famous Pukin' Dogs. They were based at the hangar right next door at Naval Air Station Oceana, so I didn't have to move. The squadron was deployed in the Persian Gulf on the USS *Eisenhower;* since we were in the middle of Operation Desert Shield, I would join them when they returned.

Being in an F-14 squadron in the 1990s was like a cross between playing a professional sport and being in a rock-and-roll band. The movie *Top Gun* didn't quite capture the arrogance and bravado of it all. The level of drunkenness and debauchery was unbelievable (and is, thankfully, no longer the standard). There were strippers in the Officers' Club every Wednesday and Friday night, and it was a big party every time. On my first day, a senior officer in the squadron told me in no uncertain terms, "This squadron is about three things: flying, fighting, and fucking, and not necessarily in that order." I told him I understood, and I did, at least about flying and fucking (sort of). But I was confused about the fighting part—I wasn't sure whether he was referring to the Soviets or some other enemy combatant, or something else. In fact, it meant going to bars on the weekends with the intention of getting into bar fights. At the annual conference for naval aviators, known as Tailhook, the debauchery reached new levels. For example, some pilots decided to make their adjacent rooms into a suite by using a chainsaw to cut through the wall. Soon after, events at a Tailhook conference would create a sexual harassment scandal that made national news and resulted in a chain reaction of investigations, firings, and policy changes. Though I had never witnessed anything as extreme as the behavior that led to the scandal, I had witnessed behavior that crossed the line, and I always wondered how this could be acceptable in the military. I didn't participate in it, but I hadn't done anything to try to stop it, either. In the long term, the policy changes have been for the better. As a result of the shake-ups caused by the scandal, women were soon allowed to fly in combat for the first time. This created a much more even playing field and advanced the careers of a few talented women pilots, some of whom would later become my astronaut colleagues. Over the course of twenty years, many more would follow.

I continued to train over the course of the next year and flew to other air stations in Key West and Nevada to practice and continue acquiring new skills. I was with the squadron when it left for its next cruise on the USS *Dwight D. Eisenhower* (we called it *Ike* for short), in September 1991. We were headed to the Red Sea, the Persian Gulf, and the fjords of Norway. I would be gone for six months, during which I would fly the F-14 nearly every other day, on combat air patrol. The Soviet Union fell apart while we were at sea, and we didn't know yet what that would mean.

One black night only a few weeks into the cruise, my RIO Ward Carroll (whom we called Mooch) and I launched in the Arabian Sea without a hitch and took up our combat air patrol position over the carrier. Our official duty was to protect our aircraft carrier's battle group from airborne threats. In other words, we were there to shoot down any bombers or fighters that might be getting anywhere near us. We also used this time to do some training. When our hour-and-a-half sortie was over and it was time to head back, I heard Mooch say, "There's land between us and the ship."

"Land?"

I was pretty sure we hadn't flown over any land. There wasn't any bad weather forecast for that day, but the horizon had completely disappeared. Then I realized that the "land" we were seeing on our radar was sand—a *haboob* in Arabic, a giant sandstorm. It had completely engulfed the region, and it was likely going to make it a very tough night behind the boat.

As we got closer to the ship and leveled off to start our final approach, visibility was awful. I heard the air traffic controller say, "Salty Dog One Oh Three, three-quarters of a mile, call the ball." He wanted me to confirm I could see the visual landing aids that would allow me to line up for landing. I looked outside and saw absolutely nothing. Then I heard the landing signal officer, who stands on the back of the ship to guide us in, say, "Paddles contact, keep it coming," meaning he could see us even if we couldn't see him. We continued our descent toward the ship.

When we were less than a quarter mile away, I could finally see the carrier. At 150 miles per hour, I had about five seconds to make corrections to line up the airplane with the centerline and adjust our altitude and speed to land in the right spot on the flight deck, right before the third arresting cable. We touched down. As usual, I went to full power, always necessary in case the landing wasn't successful and I would have to take off again instantly. I expected to feel the comforting pull of the arresting cable bringing us to a stop—but it never came.

"Bolter, bolter, bolter . . . hook skip," called the landing signal officer, the LSO. To bolter means to fail to catch the arresting wire with the airplane's tailhook. We had to immediately accelerate at full throttle in order to take off again and circle around for another attempt. Off we went back into the sandy darkness of the sky. I was frustrated because I hadn't done anything wrong—I was just unlucky. The hook had skipped over the wires. We came in again, only to experience another bolter. We came around again. Another bolter. We came around again. Now we got a wave-off, meaning our approach was so ugly they wouldn't let me try to land for fear I'd crash. Now I was seriously getting angry with myself and nervous.

The visibility was not improving, and we were running out of gas. We went around a number of times more, which only resulted in more bolters—wave-offs because we were too close to the airplane in front of us and wave-offs for performance (in other words, shitty flying on my part). Eventually we were "trick or treat," which means we either had to land this time or go get more gas. I boltered again. We were off to the tanker.

The tanker was an A-6 Intruder configured with external fuel tanks that circles overhead at three thousand feet, ready to refuel airplanes. Finding the tanker was a challenge in itself, because we were still engulfed in the sandstorm. We did a radar-only rendezvous, which was very risky, with Mooch calling out range, bearing, and closure as we approached. Once I was within twenty-five feet or so, I was able to see the tanker and join up on its wing. I extended my refueling probe, but the bumpiness of the air and the fact that I had gone around so many times in attempts to land on the ship had me thoroughly unsettled. It took me multiple attempts to make contact, during which I tried not to think about what would happen if we couldn't refuel: we would have to eject or take our chances with the barricade (a net rigged on the deck to catch the airplane), both very dangerous. Once I finally did make contact with the tanker and refueled, I headed back toward the ship.

Then I boltered, and boltered again. *I am going to do this for the rest of my life,* I thought. Eventually I put the airplane down on the deck in the right spot and felt the relief of the arresting cable's tug as we came to an abrupt stop. As I taxied forward to be chained down, I noticed that my right leg was shaking uncontrollably from the adrenaline surging through my system after all those attempts and close calls. Mooch and I made our way off the flight deck, down the dimly lit corridors smelling of jet fuel, down the ladder, and into our brightly lit ready room. The pilots burst into applause when we walked in. They'd been watching our misadventures on a monitor the whole time.

"Welcome back to *Ike.* We never thought we were going to see you fellas again."

It had been my first real "night in the barrel," and I had survived it. (There's an expression among naval aviators based on a bawdy old joke about a pilot finding sexual relief via a barrel, only to discover that his turn in the barrel was coming up.)

I laughed and accepted their congratulations.

"There are those who have," I said, "and those who will."

MY SECOND MEMORABLE bad night flying was in the Persian Gulf, on a night that was brilliantly clear at first. The moon was bright, what we called a commander's moon because the air wing's commanding officers would take advantage of it to log their night landings under easier conditions. My RIO Chuck Woodard (call sign "Gunny") and I launched that night to protect *Ike* and its battlegroup from the Iranian air force. After about an hour, the carrier's air traffic control told us we could return to the ship early. We had plenty of fuel, since we were coming back sooner than planned, so for fun, and to expedite our return, I lit the afterburner and we went supersonic. We were approaching the marshal point, an imaginary point twenty miles behind the ship where supersonic speed was not recommended. Normally I wouldn't have been going that fast, but it was such a clear night it seemed safe.

I immediately sensed I was getting behind the airplane. Even though it was clear at altitude, a layer of fog had rolled in below us, and by the time we descended through five thousand feet above the water, I was having a hard time keeping up. I started feeling rattled, sweating with my heart pounding. I was having a "helmet fire." Everything was happening too fast. I felt completely overwhelmed.

My altimeter alarm went off to warn me we were passing below five thousand feet; then it went off again to warn me we were getting even lower. It was a distraction, so I made the almost-fatal error of turning it off.

The next thing I heard was Gunny shouting, "Pull up!" Without thinking, I immediately pulled back hard on the stick, simultaneously looking over at the altimeter and vertical speed indicator. We were at eight hundred feet, dropping at four thousand feet per minute. About twelve seconds later, we would have flown into the water, becoming one of the many planes that never return to the ship. No one would have had a clue as to what had gone wrong.

With much difficulty, Gunny and I were able to gather our wits and land safely. We proceeded to my stateroom, where we cracked open a bottle of whiskey to calm our nerves and celebrate cheating death.

Leslie met me when I returned from that cruise, and I was thrilled to see her. While I had been gone, I had been deprived of so many things—the people I cared about, beer, decent food, privacy—and it was great to get them all back again. I would have the chance to experience that kind of deprivation again.

My wedding date was set for April 25, 1992, a month after I returned from the cruise. That morning when I got up and started going through the process of getting ready—taking a shower, shaving, packing my bag for the honeymoon—a strange feeling of dread loomed over me. I kept poking at it in my mind, the way your tongue keeps going to a sore tooth. This was supposed to be as happy a day as any, like the day I landed on the aircraft carrier, or the day I got my wings, or the day I graduated from college. But all I felt was this strange foreboding.

All of a sudden, as I was knotting my tie, I realized I didn't want to get married.

I cared for Leslie and enjoyed her company. But if I was being honest with myself, I wasn't marrying her because I had been moved to in my heart. I thought about the six groomsmen who were prepared to stand up with me. They were all Navy men, some from my squadron who I hadn't even known for very long. The people I had grown up with and been through trials with and who had been there for me for many years, were coming to the wedding, but they weren't in the wedding party. Without being aware of it, I had created a Navy event rather than a wedding.

I felt I had no choice but to go through with it. I wasn't going to disappoint Leslie and her family, or my own family. Mark was coming all the way from Japan, and I thought about how bewildered and annoyed he would be if he arrived to learn the wedding had been called off. By the time I was dancing with Leslie at our reception, I had managed to put all these thoughts out of my mind. Somehow it didn't feel like a permanent mistake I was making. I was only twenty-eight. I would try to make this work, but if I couldn't, I figured, I could get divorced.

I APPLIED TO U.S. Navy test pilot school in Patuxent River, Maryland, after two and a half years in the Pukin' Dogs. Usually pilots serve in a fleet squadron for four years before applying, so I didn't think I would be accepted, but I wanted to let the selection board see that my interest was serious and to familiarize myself with the application process. To my surprise, I was selected, and even better, my brother had been selected too, so we would be classmates. We started in July 1993. My biggest concern wasn't my flying, which I had become pretty confident about, but the fact that I had almost never used a personal computer. I knew I would have to get comfortable with technology, so I asked a squadron mate to help me buy one and teach me how to use it.

Leslie and I headed to Patuxent River (everyone calls it Pax River for short), only a few hours from Virginia Beach. This would be the first time in my career that I spent much time with members of the other military services. The school had U.S. Air Force pilots, Marine pilots, Army pilots; there was an Australian F-111 pilot and an Israeli helicopter pilot. Some of the people in my class would later become astronaut colleagues: Lisa Nowak, Steve Frick, Al Drew, and of course Mark. Soon after we arrived, the senior class threw a party for us, called a "You'll Be Sorry" party, warning us that we would be sorry for deciding to become test pilots because the training was so hard.

I didn't find the academic work particularly grueling, though I had to review some calculus and physics. We learned about aircraft performance, flying qualities, flight control systems, and weapons systems of the aircraft we might be testing. We also spent time familiarizing ourselves with the airplanes we would fly regularly during training. For the fixed-wing pilots like me, that meant the T-2 again, as well as the Navy version of the T-38, a much more challenging airplane. Friday nights were spent at the BOQ bar or at the home of one of my classmates. The weekends we spent doing homework.

As we got checked out in the T-38, I found landing particularly challenging, because I had gotten out of the habit of flaring an airplane—pulling back on the stick as you get closer to the ground to arrest the rate of descent prior to touchdown. When we land on a carrier we approach and land with a constant rate of descent. We also started flying other airplanes, generally with instructors or classmates who were checked out in them. This was all meant to expand our flight experience base. We also learned to write technical reports, a large part of the program. Experimenting and collecting data on a specific aspect of the airplane, then writing a detailed report on the findings, take more of a test pilot's time than actually flying airplanes.

After graduating from test pilot school in July 1994, I moved to the other side of the airfield to the Strike Aircraft Test Directorate, the Navy's test squadron for high-performance jet airplanes, located on the same base. My fighter squadron was a great fraternity, but in some ways the test pilot community was better because of its diversity. There were civilians (a group of people I hadn't previously worked with much in the military), people from different countries, different cultures, ethnicities, sexual orientations, genders, and backgrounds. I was surprised to find that diverse teams were stronger teams, each person bringing his or her own strengths and perspectives to our shared mission.

My daughter Samantha was born on October 9, 1994, in Pax River. Leslie had become more fragile and sensitive during her pregnancy, but once Samantha was born, Leslie's life revolved around her. As a mother, she was doting and full of praise. Samantha was a jolly kid, outgoing and infectiously happy.

Mark lived not far from us, and he and his wife came over often, or we would go to their home. I was part of a close-knit group of test pilots and flight test engineers, and we all enjoyed having one another over on the weekends. Leslie and I both liked having people around, which kept us from having to spend as much time alone together. My colleagues and friends all liked her, and so did their spouses. So for a while, we got along. Thanksgivings and Christmases with her family or mine were always great. I was doing the work I wanted to do and I had a family. It seemed like this was going to be my life.

In my role as test pilot I was assigned to assist in the investigation of an accident involving an F-14 that had crashed on approach to the USS *Abraham Lincoln* in a routine training mission. Lost in the accident was Kara Hultgreen, a pilot I had overlapped with in flight school. I didn't get to know her well in Beeville, but since she was one of very few women there, she was hard to miss. Shortly after we got our wings, when the Navy had just opened up combat positions to women pilots, Kara had become the first woman to qualify in the F-14. Her achievement drew a lot of attention, so it was especially distressing that she lost her life soon after, on October 25, 1994.

A video of the crash, shot from the flight deck of the aircraft carrier, showed the airplane overshooting the centerline. When Kara turned the airplane too tightly, the airflow into the left-hand engine was disturbed, causing a compressor stall—a known issue with the F-14A. (The F-14 had horrendous flying qualities in general, and the scene in *Top Gun* where Goose smashes into the canopy is one of the more accurate moments in that movie.) When Kara engaged the afterburner on the remaining engine, the rapid thrust asymmetry caused her to lose control of the aircraft. Her RIO was able to eject them both and he escaped safely, but the pilot gets ejected 0.4 seconds later, by which point the cockpit was facing the ocean. She impacted the water before her parachute opened and was killed instantly.

A new digital flight control system was being designed for the F-14 to prevent upright flat spins and crashes like Kara's. The system had taken longer to roll out than anticipated and was plagued by technical delays and cost overruns. Once our investigation of Kara's crash reached the conclusion that the new digital flight control system would likely have saved her life, the project was expedited. Soon it was ready to be tested.

Normally, first flights of new aircraft (or significant modifications to existing ones) are flown by test pilots working for the companies that manufacture them—in this case, Northrop Grumman. But because I had flown the Tomcat more frequently in the last year than anyone else, despite being relatively junior, I was chosen by our squadron commander—to everyone's surprise, including mine—to fly the first flight. The day before I was scheduled to fly, when I got into the cockpit to check out the airplane's systems, I was testing the trim button on the stick and discovered that using the button caused the flight control surface to move in the wrong direction. The lead flight test engineer, Paul Conigliaro, and I were aghast. I was supposed to take this thing into the sky the next day, and the flight control software was completely screwed up. To this day, Paul remembers my first words to the contractor responsible for the new system: "I can't tell you how much this concerns me."

When we checked the airplane again in the morning, it had been repaired—it turned out two wires had been crossed. My RIO, Bill

"Smoke" Mnich, and I rolled down the runway that morning not knowing for sure whether this airplane would leave the runway in a controlled manner. At 125 knots, I slowly pulled back the stick and we were airborne. Soon after, we were raising the landing gear and flaps. I pulled the throttles back from full afterburner and headed out over the Chesapeake Bay to begin our maneuvers. After an hour and a half of flying very slowly and methodically, expanding the flight envelope of the new system step by step, we were safely back on deck.

The F-14 was retired in 2006, and airplanes with this system never experienced another flat spin or aircraft carrier landing fatality.

9

June 21, 2015

Dreamed that Amiko arrived here on the ISS. I wasn't expecting her, so it was a pleasant surprise. She was here for work—she was setting up a public affairs event for Anton Shkaplerov—and I showed her around. It was nice to be able to welcome her to this place I've been telling her so much about. We had a conversation about whether we could both fit into a single crew quarters, and we decided we couldn't. At least, not for sleeping. She was wearing the same outfit she wore to jump out of an airplane.

BEING ALONE in the U.S. segment, I can go all day without seeing another person, unless I have reason to visit my Russian colleagues. The chatter of my crewmates is suddenly gone, and with it the chatter between each of them and the ground. I appreciate the quiet and the privacy, a rare luxury up here. I can blast music or enjoy uninterrupted silence. I keep CNN on all day, at least when the satellites are lined up, to keep me company.

I do sometimes miss having another person to talk to, even if it's just to complain about the challenging work schedule or to talk about what's on the news. On a more practical level, I often miss being able to get a bit of help now and then. Many of the tasks on my schedule are doable by one person but would be much easier with another pair of hands at key moments. My workdays are longer when I do everything alone. The cosmonauts would drop everything to help if I needed them to, but they have their own work, and the delicate exchange of labor, resources, and money between our two space agencies is complex. I don't want to complicate it further by asking for free help.

Today is Gennady's birthday, and we have a special dinner in his honor. I give him the gift I remembered to pack: a ball cap with embroidered U.S. Navy pilot's wings. Today is also Father's Day, so we wind up talking about our children. Gennady has three daughters—two now grown and a twelve-year-old like Charlotte, along with a granddaughter close in age to his youngest daughter. He says he has regrets about missing his daughters' childhoods because he was so focused on his career. He says he's a much different father now than when they were younger. We both say we are looking forward to spending more time with our kids when we get back.

After we say our good nights and I go back to my CQ, I find an email from my ex-wife Leslie, which is unusual. She generally doesn't deal with me directly. She wanted to let me know that she had heard from Charlotte's teacher. A few days ago, Charlotte's class was playing a game, and she was first to choose her teammate. Charlotte could easily have chosen one of her friends, but instead she chose a classmate who is developmentally challenged and has never been chosen first for anything. The teacher was so touched, she created a special award for Charlotte for always doing the right thing. Leslie's email makes me feel both closer to Earth and farther away at the same time. It nearly brings a tear to my eye.

I WAKE UP early in the morning, six a.m., and float out of my CQ, through the lab and Node 1, turning on lights as I go. I turn right, into Node 3, where I go into the WHC. I don't start it up, though—today is a science sample collection day. The process of urination is going to be even more complicated than usual. I grab a urine collection bag, clear plastic with a condom attached to one end. I put the condom on, then wrap it in mesh bandages to prevent leaks. As I urinate, I have to push with enough force to unseat the valve on the bag to allow the urine to flow in—without the valve there, of course, it would just come floating back out. But it's hard to push with enough force to open the valve without pushing so hard the urine leaks from the condom—and this is exactly what happens. Urine soaks the gauze, then quickly forms droplets that float out to the walls. I'll have to clean them up later. After I finish peeing, I remove the condom while trying not to liberate more urine. I use sample tubes with plungers to draw out three samples, initial them, mark them with the date and time, and scan their barcodes into the system. Then I head down to the Japanese module to put the tubes into one of the freezers. I will go through this process again and again, every time I urinate for the next twenty-four hours.

With the pee sample done, I head into Columbus for my blood draw. Like most astronauts on ISS, I know how to draw my own blood. At first I told the instructors in Houston that I wouldn't be able to stick a needle in my own vein, but with some help I agreed to give it a try and quickly got the hang of it. Gennady joins me in Columbus to help, right on time, though I told him last night he didn't need to. I clean the site on my right arm, which I've found to be a better vein. Using my left hand, I pierce the skin and slip the needle in. There is a brief flash of red in the tube holder, an indication that I hit the vein, but when I connect the vacuum tube, there's no blood. I must have gone right through. Having ruined that one for today, I will have to try again on the left side. Because this is my only remaining arm, I suggest Gennady give it a try for me.

Gennady grabs another butterfly needle and connects it to the tube holder. After cleaning the site on my left arm, he takes aim and slides the needle perfectly into the vein. But the needle isn't properly connected to the tube holder, so blood escapes, flowing out into globs in the air that wobble and then resolve themselves into crimson spheres, traveling out in every direction. Gennady quickly reseats the connection while I reach out to grab some of the blood spheres with my hand before they can float farther away. The ones I missed I'll have to track down and clean up later. Luckily, I'm mostly alone on the U.S. segment, so no one will encounter a gory surprise before I can get to it.

Gennady changes out the tubes over and over until he's drawn ten tubes of blood. I thank him for his help, and he goes back to the service module to have breakfast. I put the tubes in the centrifuge for half an hour, then put them in the freezer along with the other samples.

Later in the day, I will take a fecal sample; tomorrow, saliva and skin. I will go through this whole process every few weeks for the rest of the year.

Within the past week I've developed a badly infected ingrown toenail on my left big toe. Almost every moment of the day, unless I'm sleeping, I have one or both feet hooked around a handrail to hold me steady, so big toes are extremely important. I can't afford for this guy to be out of commission. I'm treating it with topical antibiotics—we have a full pharmacy up here—and monitoring it closely.

The CO_2 is much better now that I'm the only one exhaling on this side of the ISS. My headaches and congestion have largely cleared up, and I notice a difference in my mood and cognition. I'm appreciating this break from the symptoms while I can. At the same time, I'm concerned because the ground will probably act as though there is no problem now. Then the next crew will get here and we will start the whole cycle all over again.

One of the nice things about living in space is that exercise is part of your job, not something you have to fit in before or after work. (Of course, that's also one of the bad things about it: there are no excuses.) If I don't exercise six days a week for at least a couple of hours a day, my bones will lose significant mass—1 percent each month. We've had two astronauts break their hips after long-duration spaceflights, and since the risk of death after hip fracture increases with age, bone loss is one of the biggest dangers my year in space will pose to my future health. Even with all this exercise, I will lose some bone mass, and it's suspected that bone structure changes permanently after long-term spaceflight (this is one of the many medical questions Misha's and my year will help to answer). Our bodies are smart about getting rid of what's not needed, and my body has started to notice that my bones are not needed in zero gravity. Not having to support our weight, we lose muscle as well. Sometimes I reflect that future generations may live their whole lives in space, and they won't need their bones at all. They will be able to live as invertebrates. But I plan to return to Earth, so I must work out six days a week.

When it's workout time on my schedule, I float into the PMM, a windowless module we use as a large closet, to change into shorts, socks, and a shirt. The PMM always reminds me of my grandparents' basement—it's dark, dingy, and has random stuff everywhere. My workout clothes are getting a bit fragrant because I've been using them for a couple of weeks—there is no laundry up here, so we wear clothes for as long as we can stand, then throw them out. I struggle to find something to hook my feet onto while I change. The clothes are still moist from yesterday's exercise, making changing unpleasant.

I head into Node 3 and make my way to the treadmill. On the ceiling is a strap that holds a pair of shoes, a harness, and a heart monitor for each of us. I grab my running shoes and put them on, then I step onto the treadmill, which is mounted on the "wall" with respect to most of the other equipment.

I put my harness on, buckle it at the waist and chest, and clip into the bungee system that's attached to the treadmill. This holds me in place as I run—without the harness, I would go flying off the treadmill with my first step. We can adjust the tension to control the perceived weight at which we're running, though we can't run at our normal body weight, as the pressure on our hips and shoulders would be too painful. I set up the laptop in front of me and start an episode of *Game of Thrones*. I deliberately avoided watching the series when it first aired and people were talking about it because I knew I would need some good escapist entertainment this year. Now I'm watching the whole series for the second time.

In some ways our treadmill is like the one you might find in a gym on Earth, but it's mounted into its own unique vibration isolation system. The forces created by the runner pounding away could be surprisingly dangerous—an oscillation at the wrong frequency could tear the space station apart. On *Mir,* Russian mission control once had to ask American astronaut Shannon Lucid to run at a different pace or risk damaging the space station. On his first flight, cosmonaut Oleg Kononenko, who will join us up here soon along with Kjell and Kimiya, created a potentially dangerous oscillation just by absentmindedly floating up and down a few inches, his feet gently pushing against the floor and a bungee cord.

I control the treadmill using software on the laptop, starting off slowly then gradually ramping up. I enjoy the daily exercise, but it's hard on my joints. Some days the pain in my knees and feet is almost unbearable, though today it's not too bad. I ramp up to my maximum speed. When I sweat, the liquid builds up on my bare head like water on a newly waxed car. I wipe it away with my two-week-old sweat towel. Once in a while other people come floating through, positioned perpendicular to me. It's hard to sneak by the person on the treadmill without distracting them or, worse, hitting or kicking them, especially for people who are new on station. It takes some getting used to, seeing someone running on the wall.

While I'm running, Gennady comes by to check on something. There are some shit cans temporarily stored in a big bag on the floor of Node 1, waiting to go out on the outgoing Progress with the rest of the garbage, and Gennady had noticed they were smelling a bit. He checks one of the lids to make sure it's sealed properly, only to accidentally liberate a cloud of toxic gas that nearly knocks me off the treadmill. It makes me think of the Monty Python sketch where everyone triggers one another to throw up. The entire U.S. segment smells wretched for a while, but I'm impressed by how quickly the system filters the air.

"As soon as I get back to Earth," Gennady mutters in Russian, "I am going on a vacation."

Soon after he leaves, I hear the voice of mission control.

"Station, Houston on Space to Ground Two. We are privatizing the space-to-ground channel. The flight director needs to speak to you."

We are privatizing. These are words that make any astronaut's blood freeze. They mean something bad has happened. I bring the treadmill to a stop, unhook myself, and grab the mic to talk to Houston.

The last time I heard "we are privatizing" was when SpaceX blew up. The time before that, my daughter Samantha was having a personal crisis. And of course on my last mission, the news that we were privatizing came when my sister-in-law was shot. I wait anxiously to find out what has gone wrong.

I hear the capcom on duty, Jay Marschke, refer to the trajectory operations officer (TOPO). For a moment, I'm relieved; at least it has nothing to do with my family.

"This is a red late-notice conjunction," Jay says, "with a closest point of approach within a sphere of uncertainty."

"Roger," I say into the microphone. Then I make sure the microphone is off before I say what I really think about this, which is, "*Fuck.*"

A "conjunction" is a collision—a piece of space junk is headed our way, in this case an old Russian satellite. "Late notice" means we didn't see it coming or that we miscalculated its trajectory, and "red" means it's going to get dangerously close—we just don't know how close. The "sphere of uncertainty" refers to the area it could pass through, a sphere with a radius of one mile. Because the impact could depressurize the station, letting our air out and killing us all, we will have to head to the Soyuz and use it as a possible lifeboat. If the debris streaking toward us collides with us, we will likely all be dead in two hours.

"How about relative velocity?" I ask. "Any idea?"

"Closing velocity of fourteen kilometers per second," comes the answer.

"Copy," I say into my headset. ("Fuck," I say, again, to myself.) This is the worst possible answer to my question. If the satellite were in an orbit similar to ours, the closing speed might be as low as a few hundred miles per hour—a devastating speed for a car crash, but a best-case scenario for a space crash. Instead, the space station is traveling in one direction at 17,500 miles per hour, and the space junk is traveling at the same speed in the exact opposite direction; a 35,000-mile-per-hour closing rate—twenty times faster than a bullet from a gun. If the satellite hits, the resulting destruction would be much worse than what happens in the movie *Gravity.*

With six hours' notice, the space station can move itself out of the way of oncoming orbital debris. The Air Force tracks the position and trajectory of thousands of objects in orbit—mostly old satellites, whole or in chunks. As with everything else, NASA has an abbreviation for these adjustments: PDAM, or predetermined debris avoidance maneuvers, which means firing the station's engines to adjust its orbit. We've had two of this type since I've been up here. Today, however, is different. With only two hours' notice, a PDAM will not be possible.

Mission control directs me to close and check all the hatches on the U.S. segment of the space station. I trained to do this in my preparations for this mission, and I run through the procedure in my mind in order to complete the steps properly and—most important—quickly. Even the hatches that were already closed need to be checked, like the unused berthing ports for visiting vehicles. With the hatches closed, if one module is hit, the others might survive—or at least their contents won't be sucked out into the vacuum of space. There are eighteen hatches on the U.S. segment that must be closed or checked. While I'm working through the hatches as efficiently as I can, I get a call from mission control.

"Scott, Misha, it's time to get ready for your event with WDRB in Louisville, Kentucky."

"What?" I ask, incredulous. "Is there really time to be doing this?"

Misha shows up in the U.S. lab for our joint public affairs event, as he always does, with no time to spare but right on time.

"Public affairs events can't be canceled," comes the answer. The anchors want to ask us about watching the Kentucky Derby, which was almost two months ago. This is insane.

"Are they fucking kidding?" I say to Misha. He shakes his head in response. This is a bad decision, but it's also not a great time to get into an argument with the ground.

Misha and I get into position in front of the camera with the handheld microphone.

"Station, Houston, are you ready for the event?" Jay asks.

"We are ready for the event," I answer, struggling to keep the annoyance out of my voice. We spend the next five minutes answering questions about what we think of the probe that just reached Pluto, what landmark we may be passing over, and whether we got to watch the

Kentucky Derby back in May. This kind of interview is part of our job, but today we can't help but grit our teeth.

When we are asked about maneuvering in weightlessness, we turn somersaults for the Louisville viewers before signing off, still feeling pissed off that we had to waste our time this way, given the magnitude of the situation we are in. There is danger in becoming too complacent about the reality of life on an orbiting space station, and the decision to go ahead with this interview is, to me, clearly a symptom of that.

As soon as the cameras are off, I get back to checking that the hatches are closed. Luckily there are no serious issues with any of them—I don't have the time to fix any problems. I collect the items from the U.S. segment that we will need most if a collision destroys that part of the station: the defibrillator, the advanced life support medical kit, my iPad with important procedures on it, my iPod, and a bag of personal items. I also make sure I have my thumb drive of images and videos from Amiko that I wouldn't want to lose track of. By the time I have gathered all my important items, we have about twenty minutes to spare before potential impact.

I go to the Russian segment, where I see that the cosmonauts have not bothered with closing their hatches. They think closing the hatches is a waste of time, and they have a point. The two most likely scenarios are that the satellite will miss us, in which case closing the hatches will have been pointless, or it will hit us, in which case the station will be vaporized in an instant, and it won't make a bit of difference if the hatches are open or closed. It is incredibly unlikely that one module could be hit and the others survive intact, but just in case, mission control has me spend more than two hours preparing for that eventuality; the Russian approach is to say fuck it and spend what might be their last twenty minutes having lunch. I reach my crewmates in time to join them for a small can of Appetizing Appetizer.

Ten minutes before potential impact we make our way to the Soyuz, which Gennady has prepared for flight in case we have to detach from the station. It's orbital night now and dark in the Soyuz as we each slide into our seats. It's cramped and cold and loud.

"You know," Gennady says, "it will really suck if we get hit by this satellite."

"*Da,*" Misha agrees. "Will suck."

Only four other times in fifteen years have crews had to shelter in place as we are now. I can hear our breathing over the sounds of the fans stirring the air inside the Soyuz. I don't think any one of us is actually fearful. We've all been in risky situations before. We do talk, though, about the size and velocity of the piece of space junk coming toward us. We all agree that it's a potentially disastrous scenario.

Misha stares out the window. I remind him that he won't be able to see the satellite coming toward us—it will be going way too fast for the human eye to perceive, and besides, it's dark outside. He keeps looking anyway, and soon I'm looking out my window too. The clock counts down. Once the time gets down to seconds, I feel myself tensing, starting to grimace. We wait. Then . . . nothing. Thirty seconds go by. We look at one another with a last heartbeat of anticipation of disaster. Then our grimaces slowly turn into expressions of relief.

"Moscow, are we still waiting?" Gennady asks.

"Gennady Ivanovich, that's it," Moscow mission control responds. "The moment has passed. It is safe; you can go back to work now."

We float out of the Soyuz one by one, Gennady and Misha finish lunch, and then I spend most of the day opening all the hatches.

Later, as I reflect on the situation, I realize that if the satellite had in fact hit us, we probably wouldn't even have known it. When an aircraft flies into a mountain in bad weather, at five hundred miles per hour, there is little left to tell the story of what went wrong: this crash would have taken place at a speed seventy times that. When I used to work on investigations of aircraft mishaps as a Navy test pilot, I would sometimes reflect that a crew might never have known that anything had gone wrong. Misha, Gennady, and I would have gone from grumbling to one another in our cold Soyuz to being blasted in a million directions as diffused atoms, all in the space of a millisecond. Our neurological systems would not even have had time to process the incoming data into conscious thought. The energy involved in a collision between two large objects at 35,000 miles per hour would be similar to that of a nuclear bomb. I think of that time I almost flew an F-14 into the water and would have disappeared without a trace.

I don't know whether this comforts me or disturbs me.

IN ELEVEN DAYS, a new crew will arrive. I try not to think about how much more time I will have up here, since I know that will only make it harder. But my year in space will divide itself neatly into four expeditions of three months each, and when Kjell, Kimiya, and Oleg arrive, that will mark the passage of only one-quarter of my time here.

10

July 24, 2015

Dreamed I was on Earth, visiting New York with Amiko. We got into a taxi, and I noticed Amiko was carrying a cage with some huge spiders in it, big ones like the goliath bird-eating tarantula named Skittles that I bought Samantha for her birthday a few years ago. Our taxi driver was named Jenny, and she told us she was a postal worker moonlighting by driving a cab—in fact, she had some of the mail she was supposed to be delivering in the trunk. I got into an argument with Jenny about something, and she kicked us out of the car and drove off, with Amiko's spiders still in the backseat. I ran after the car and got the spiders back for Amiko, then laughed when I noticed Jenny now had a flat tire.

TODAY the Expedition 44 crew arrived. Their launch was a relief after the recent Progress failures, and docking went off without a hitch. When we opened the hatch and the new guys came floating through, looking dazed as baby birds right out of the shell, I was reminded of the day I passed through the same hatch in my Captain America suit, Misha and me fitting through the opening together like a set of conjoined twins. It feels like that was years ago. The days are going by quickly, but the weeks crawl.

The three new guys will need a lot of help acclimating to the environment, getting settled, and learning to do the work. For experienced astronauts serving on ISS for the first time, the adjustment period is longer than for those who have lived here before; for first-time space travelers, like Kjell Lindgren and Kimiya Yui, it's longer still. (This is Oleg Kononenko's third time in space.) I trained a year for each of my flights on the space shuttle, preparing in detail for each day's activities on a two-week mission. In the ISS era, with such a large spacecraft and much longer missions, our training is more generic. We don't know exactly what we're going to be doing from day to day. It's much more challenging, and the biggest challenge is at the start of the mission.

More than two-thirds of space travelers suffer from some degree of space motion sickness, sometimes debilitating, and there isn't much to be done but wait it out. Kjell and Kimiya both feel pretty bad on day one, and they will be nauseous and only marginally functional until their bodies adjust to the disorientation of zero g. Until they fully adapt, they will be as clumsy and tentative as babies just learning to walk. They will need help with the simplest things; even moving from one module to another without knocking shit off the walls is a challenge. They will need help talking to the ground, preparing food, using the bathroom. Even the process of throwing up requires help initially. It will take them four to six weeks to feel fully normal.

Soon after the new guys come floating through the hatch, we have a quick videoconference with the ground so they can greet their families, still in Baikonur. Most of the questions from the ground can be answered by just saying, "I'm fine. It was the ride of my life." Misha helpfully floats an apple and an orange behind Kimiya as a visual aid while he talks.

I know Kjell and Kimiya won't sleep well their first night here. In the middle of the night, I get up to use the bathroom and find Kjell going through bags of stuff in one of the storage modules.

"Hey, what are you looking for?" I ask him. It's close to impossible to find anything even with the lights on, and out of courtesy Kjell has left them off.

"To tell you the truth, I'm looking for more puke bags," Kjell says. "I'm out."

"There's got to be some more here somewhere," I say. I look in the few places that seem most likely, then search in the computer inventory management system. I ask Houston where I should be looking. After a minute, they say we don't have a stash of puke bags on board. We don't include them among supplies sent up to the station because the Russians used to bring them on the Soyuz.

"We'll improvise something," I assure Kjell. As with everything else up here, vomit has a tendency to go everywhere, so there has to be a way for the bag to absorb it and hold it in place. It's also nice to be able to wipe off your face, since surface tension holds liquids on your skin when they can't drip off due to gravity.

Rooting through our supplies, I invent a new puke bag for Kjell made out of a ziplock bag lined with maxi pads. It works.

For much of what Kjell and Kimiya do on their second day, they need me floating at their elbows, talking them through the procedures, offering help in learning to maneuver in zero g. Kjell's first task is to inventory the contents of a bag of spare parts that came up with them on their Soyuz and then stow them on ISS. On Earth, it would be a simple task—you could put the bag down on the floor, take everything out, and check off each item on a list as you put it back in. In space, as Kjell quickly learns, the moment you open the bag, objects jump out at you and start drifting away. Just getting everything back under control can take up all of the time that was allotted for the job.

Working through this together is time-consuming, but it will be worth it in the long run. I'm teaching Kjell general techniques he can use throughout his time here—for instance, the importance of putting things away in the right places. I tell Kjell he can keep the contents of a container from leaping out at him when he opens it by slowly spinning in place while holding the bag. Centrifugal force pushes the contents toward the bottom of the bag and holds them there as long as you keep spinning. Organizing the parts being inventoried is a bit trickier, but I show Kjell how to use a mesh bag to hold the objects that would otherwise be floating all over the lab and perhaps hiding themselves. Then he can move each item from the mesh bag back into the original bag as he accounts for it. For small or delicate objects, I show him how to lay out long pieces of duct tape faceup on the wall, bisected by shorter pieces facedown to hold the long one in place. Then he can stick items on the tape, keeping them from wandering off. There are patches of Velcro strategically placed on the walls, and new items often come up with Velcro dots affixed. It's hard to express how much easier this makes life; when a certain number of new items arrive *without* Velcro dots, I express annoyance that probably seems out of proportion to those on the ground. But every object that arrives without Velcro on it threatens to rob me of time, patience, and ingenuity, all of which are sometimes in short supply.

KJELL HAS a great attitude so far and seems enthusiastic about everything he approaches, even though he looks a bit pale, with dark circles under his eyes. Every once in a while, he gets a distracted look, then excuses himself to throw up. The first few days in space can make anyone cranky, but Kjell doesn't seem to have forgotten for one second that he is living his boyhood dream, and his positive attitude is contagious.

Kjell was born in Taiwan, to a Chinese mother and a Swedish American father. They moved to the American Midwest and then to England, where Kjell spent most of his childhood. He grew up wanting to be an astronaut, and when he was only eleven he wrote to the Air Force Academy asking for an application. When he applied as a high school senior, he was admitted and did well there. His plan had been similar to mine: become a pilot, fly jets for the military, become a test pilot, then apply to NASA and fly the space shuttle.

But after Kjell finished his degree at the academy and started flight school, a flight surgeon diagnosed him with asthma, a disqualifying condition. Kjell hadn't experienced any symptoms, but the flight surgeon's verdict was absolute. It looked like Kjell would never get to fly military aircraft. Forging a new life plan, he became a researcher studying the cardiovascular effects of spaceflight, then earned a medical degree. He completed residencies in emergency and aerospace medicine, then earned a master's in public health. He went to work for NASA as a flight surgeon, looking after astronauts preparing to go to space.

Some of Kjell's new flight surgeon colleagues were surprised or even skeptical about why he had been grounded when they heard his story. He still had not experienced any symptoms of asthma, never taken asthma medication, and was an avid runner and in great health. Some of his colleagues at the Johnson Space Center pointed out to him that while he might have been disqualified from military aviation, NASA went by its own rules. They encouraged him to apply when there was a call for new astronauts, and he did. When he was examined, no trace of asthma was found. Kjell was accepted into the astronaut corps in 2009.

I met Kjell for the first time in Star City, when he was a flight surgeon and I was training for Expedition 25/26. He is sincere and enthusiastic without ever seeming fake or calculating. He is a little on the tall side for an astronaut, with a military haircut and demeanor, but with a perpetual smile. Kjell is religious but is tolerant and respectful of other people's beliefs; he is one of the most positive people I have ever met.

Kimiya Yui has a background similar to the one Kjell thought he was going to have. He went to Japan's military academy and joined the Japan Air Self-Defense Force. He flew the F-15 fighter jet and then became a test pilot. Like Kjell, he joined the astronaut corps in 2009, the first class to join NASA knowing they would never get to fly on the space shuttle. Kimiya is an outstanding pilot. He is also one of the hardest-working people I've ever known. It's difficult enough to learn the systems of the space station, the inner workings of the Soyuz, and a foreign language, all at once—Kimiya learned *two* foreign languages (Russian and English).

Kimiya is one of seven active Japanese astronauts (there are approximately forty-five active American astronauts and sixteen representing the European Space Agency). When I first got to know him in training, he seemed very formal—though I had no way to measure that, never having been assigned with a Japanese astronaut before. He would call me "Kelly-san," a formal (though not the most formal) way to address another person in Japan. When I kept trying to get him to just call me "Scott," he started calling me "Scott-san"; eventually, he stopped calling me anything at all. Kimiya understands that Americans value informality and equality—at least in our interactions—and he tries to meet us halfway, even if it makes him uncomfortable. Yesterday, while using the water dispenser, he saw me floating toward him out of the corner of his eye. He greeted me and moved out of the way, acting as though he was busy doing something else. But as soon as I finished getting my water and floated away, I saw him going back to the dispenser to finish filling his water bag.

Oleg Kononenko is a seasoned cosmonaut and a brilliant and rigorous engineer. He is a quiet and thoughtful person, consistently reliable. He is the same age as me and has a pair of twins the same age as Charlotte, a boy and a girl.

Kjell and Kimiya got to know each other well while training for this mission, including a wilderness course through the National Outdoor Leadership School, meant to put us into high-pressure situations, somewhat like the ones we might face in space. I wasn't on that training course, since we weren't originally supposed to be on the same crew, so we'll have to get to know one another up here. In the fall, I'm going to perform two spacewalks with Kjell, and our lives will depend on us working together.

TODAY, Kjell, Kimiya, and I are all taking our blood, then separating it in our state-of-the-art centrifuge before storing it for eventual return to Earth.

The Russians are taking blood today too, and I go to their service module to pick up some of the samples they asked us to store in our freezer. As soon as I pass through the hatch to the Russian segment, the modules are smaller and more cluttered, the equipment is louder, and the ambient light is yellower. But this time it's worse: the Russians are starting up their centrifuge as I arrive, and it sounds like a chainsaw. All three cosmonauts laugh when they see my reaction.

"Can you believe this?" Gennady asks, gesturing at the centrifuge, then at his ears. "Fucking *blya*."

"That thing sounds like it's about to blow," I say, and the Russians laugh some more. If their centrifuge were to come apart, it could take the hull of the service module with it, and we would all die.

I float back to the U.S. segment, shaking my head, my ears still ringing. I feel awful from my brief exposure to the noise—like nails on a chalkboard, but much worse.

This is just another example of the differences between our countries' approaches to equipping the station. The Russian space agency's goal is always to get the job done as cheaply and efficiently as possible, and I have to admit their cost-saving solutions for some problems can be impressive. The Soyuz that gets us up and down from space is a great example of this: it's cheap, simple, and reliable. But ultimately, because the Russian hardware is unsophisticated, they are limited in the science they are able to get done. And of course at times like today, I worry about the safety of their equipment.

Kjell and Kimiya are growing used to the strangely sterile life up here. But at least now we have some plants: we have begun an experiment in the European module growing lettuce in a system that uses LED lights to bathe a plant "pillow" of control-release fertilizer. We are learning more about the challenges of growing food in space, which will be important if humans are to make a journey to Mars.

Because I've already spent so much time up here, I'm able to tune in to the subtleties of the station. I can feel a slight temperature change from one side of a module to the other. I can feel the vibration in a handrail changing slightly. The sounds of the equipment—always whirring, humming, buzzing—vary almost imperceptibly. I'll stop Kjell or Kimiya floating through and ask, "Do you hear that whooshing sound?" Often they won't have noticed it until after I've pointed it out. This hypervigilance isn't entirely a good feeling. It's another symptom of not being able to detach and shut down, of never really being off the clock. But it might keep us safer—if something were to start to go wrong I might have an early indication of it.

I recently noticed that my brain has made a transition to living in zero g—I can now see things in all orientations. If I'm "upside down" relative to the module I'm in, instead of the environment looking foreign and disorienting, as it would if you stood on your head in an equipment-packed laboratory on Earth, now I immediately recognize where I am and can find whatever I need. This is a transition I never made last time, even after 159 days in space. This may have to do with the six weeks I spent by myself on the U.S. segment—without seeing another astronaut oriented in the normal "upright" way, I was maybe better able to adapt. Or it may be that this is a transition that takes the human brain more than six months to complete. If so, I may have found one of the answers, albeit a small one, that Misha and I are here in search of.

I've been noticing that Misha has a different philosophy of pacing himself through the year than I do—he often announces the exact number of days we still have to go, which bugs the shit out of me, but I keep that to myself. I prefer to count up rather than counting down, as if the days are something valuable I'm collecting.

TODAY I AM doing a Twitter chat, answering questions from followers "live." Because my internet connection can be slow, I'm dictating my answers to Amiko and another public affairs person, and they are typing them into Twitter almost in real time. I'm answering the usual questions about food, exercise, and the view of Earth when I receive a tweet from a user with the handle @POTUS44, President Obama.

He writes, "Hey @StationCDRKelly, loving the photos. Do you ever look out the window and just freak out?"

Amiko and I share a moment of being pleased that the president is following my mission. I think for a moment, then ask Amiko to type a reply: "I don't freak out about anything, Mr. President, except getting a Twitter question from you."

It's a great Twitter moment, unplanned and unscripted, and it gets thousands of likes and retweets. Not long after, a reply appears from Buzz Aldrin: "He's 249 miles above the earth. Piece of cake. Neil, Mike & I went 239,000 miles to the moon. #Apollo11."

There is no good way to engage in a Twitter debate with an American hero, so I don't. In my mind, I reflect on the fact that the crew of Apollo 11 spent eight days in space, traveling half a million miles; by the time I'm done I will have spent a total of 520 days in space and will have traveled over two hundred million miles, the equivalent of going to Mars and back. Only later, when the Twitter chat is over, do I have the chance to reflect that I just experienced being trolled, in space, by the second man on the moon, while also engaging in a Twitter conversation with the president.

A few days later, it's time to harvest the lettuce we've been growing. Kjell, Kimiya, and I gather in the European module to eat it with some oil and vinegar, and it's surprisingly good. This is the first time American astronauts have eaten a crop grown in space, though the Russians have grown and eaten leafy greens on previous missions. As often happens, the public reaction to the space lettuce surprises me—people seem to be fascinated by the idea of growing and eating plants in orbit, while at the same time Misha and Gennady are outside doing a spacewalk that gets no attention in the United States whatsoever. Kimiya confides in me later that he had to force himself to eat the lettuce for the camera—he grew up on a lettuce farm, and in the summer he had to get up in the middle of the night to harvest it, so since then he's hated lettuce.

That evening, we bring the Russians some lettuce to sample for Friday dinner. The main topic of discussion is the Soyuz that will be coming up soon, bringing our total to nine. We talk about the new guys—Sergey, Andy, and Aidyn—and I mention that I've never met Aidyn and don't even have an idea of what he looks like. That's incredibly unusual: before flying in space with someone, even someone from another country, you normally train with him or her, if only a little.

Gennady offers to show me a picture. I decide it will be fun if I have no idea what he looks like until he comes floating through the hatch. Oleg and Gennady agree this will be entertaining.

At about one a.m. I'm awakened from a dead sleep by the failure of one of our power channels. The outage takes down half the power in Node 3, which is where we keep most of our environmental control equipment—our O_2 generator, one of the Seedras (the one I hate), and all the equipment that processes our urine into water, including the toilet itself. It takes a couple of hours working with the ground to get things under control, at which point I tell the rest of the crew to go back to sleep. I stay up another hour and a half myself as the ground attempts to regain ventilation and smoke detection capability. The culprit turned out to be a power regulator way out on the truss. By the time I get back to my CQ, I know I'll only get a couple of hours of sleep at most.

Later, when I talk to Amiko, she tells me she was in mission control when the power went down. It hadn't occurred to me that she might have been there on console, watching the displays light up like a Christmas tree with our power failure. We haven't talked much about the fact that her job at NASA could potentially put her in the strange position of watching live while some disaster threatens her partner's life.

"I bet that was scary," I say.

"Yeah, it was a bit scary," she says. "But I stayed on console until I could see that everything was okay." She tells me that soon after, the lead flight director Mike Lammers came by her console to see how she was doing. Mike was also the lead flight director for the second half of my previous mission to ISS; he is someone I trust, the person I always think of as my space station flight director. He has also set himself apart from many of the other flight directors in mission control by checking on Amiko, congratulating her on the accomplishments of the mission so far, and supporting her as my partner.

The next day, I talk to Charlotte on the phone. She's still not much of a phone person, and as usual she seems to be distracted by something, maybe the TV. She's never rude or unfriendly, but her answers are brief and vague. After a while I run out of topics and start wrapping up the conversation.

"Okay, I should get going," I say. "Just wanted to check to see how you were doing."

I expect her to answer with a good-bye, but instead there is a pause. For a second, I wonder whether we have lost our connection.

"Tell me how *you're* doing, Dad," Charlotte says. She suddenly seems to be giving me her full attention, speaking to me in a more adult tone.

I tell her about the power failure last night and how we dealt with it. She sounds interested and asks questions. I tell her more about my new crewmates and how they're settling in. I tell her about some of the experiments I've been working on and about how the clouds and airplane contrails over Europe looked while I was drinking my morning coffee. By the time we get off the phone, I feel as though I've observed a milestone of growth in Charlotte, like watching her take her first step or say her first word. She seems as though she's matured by years within the space of a short phone call. It's another milestone I've spent away from Earth.

ON THE WEEKENDS, I don't set an alarm, letting myself wake up naturally, maybe an hour later than usual. One Sunday morning in mid-August as I'm slowly waking up, I start to become aware of a welcome sound I haven't heard for many years. Maybe I'm dreaming about the weekend mornings when I was growing up in New Jersey, when the bagpipers would play at the nearby high school football field. The sound would drift into my room and wake me in a pleasant way, unlike the sounds of my parents fighting that sometimes woke me in the night.

Fully awake, I know I'm in my sleeping bag in my CQ, not in my childhood bed on Greenwood Avenue. But I'm still sure that I'm hearing bagpipes: "Amazing Grace." I make my way out of Node 2 and follow the sound to find an unexpected sight: Kjell floating at the far end of the Japanese module, playing the bagpipes. Astronauts have been bringing instruments to space for decades—at least as far back as 1965, when astronauts played "Jingle Bells" on the harmonica. As far as I know, Kjell is the first bagpiper in space.

"Sorry, did I wake you?" says Kjell.

"No, it's great," I said. "Play anytime you like."

Today, Gennady, Misha, and I are moving a Soyuz, the one Gennady will go home in, to the aft of ISS in a complex shell game designed to most efficiently utilize the docking ports. Gennady could move the Soyuz by himself, but Misha and I must come along for the ride because this Soyuz is our lifeboat, and once it undocks it's never guaranteed we will be able to get back aboard the station.

On Earth, moving the Soyuz would be as simple as reparking a car. Up here, as we get into our Sokol suits, we jokingly refer to this brief journey as our summer vacation away from the ISS. Even though we are only away from the station for twenty-five minutes, the whole procedure takes several hours with all of the preparations. As the right seater, I don't have much to do, so I bring my iPod to listen to my classical music playlist including Mozart, Beethoven, Tchaikovsky, Strauss, and Samuel Barber's *Adagio for Strings*. I had almost forgotten how uncomfortable this position is on my knees—I'm not looking forward to doing this again seven months from now when we leave the station for the last time.

When we push away from the station and start the fly-around, I find it strange to see the station from the outside again. It's been five months since I've been outside. Even though it's in jest that we've been calling this our "vacay," it is actually good to get away. And like earth-bound vacations, this one feels too short and leaves me somehow feeling more tired than when I left.

11

ONE AFTERNOON in early 1995, I was in my cubicle at the test squadron, in a trailer adjacent to a row of World War II–era hangars and the flight line where F-14 Tomcats and F/A-18 Hornets sat idle ready for flight, when I noticed that one of my colleagues had a big stack of papers on his desk. I asked him what he was doing.

"I'm filling out my astronaut application," he said.

Of course, I had been aiming to fill out an astronaut application someday, but I'd assumed I wasn't ready yet, that I wouldn't be for another ten years or so. I was only a little more than a year out of test pilot school, just thirty-one years old, a bit on the young and inexperienced side for a pilot astronaut. I also didn't have a master's degree yet, which I thought was a requirement. But I asked my colleague whether I could take a look at his application. I was curious about what was involved, and I was especially curious about why his stack of papers was so thick. When I paged through it, I saw that NASA was looking for a lot of the kinds of information you would expect: transcripts, letters of recommendation, a detailed list of job responsibilities to date. I also noticed that he had included everything he had ever done in his life. This guy was one of the best qualified among us.

Looking over his application, I had an idea: Why not apply and be rejected? It would give me the opportunity to find out what the process was like, and rejection wouldn't harm my chances in the future. I decided to take a different approach from what my colleague had done. I would include only what seemed really important. If my application was brief and concise, maybe the person reading it could take in all the information and be left with a clear sense of who I was. This minimalist approach was also appealing to me because the deadline was fast approaching.

I filled out the application for federal employment and submitted it on time. Months later, my colleague whose application I had first seen shared the news that he'd been called for an interview with NASA, in the first week of interviews. The conventional wisdom at the time was that NASA called their top choices first, and those interviewed in the first round had by far the best chance of being selected. I congratulated him and figured I would never hear anything.

A few weeks later, Mark and his wife were having dinner with Leslie and me at our house. Halfway through the meal, Mark announced that he had also been called for an astronaut interview.

"That's awesome, congratulations," I said. And I meant it. I felt he truly deserved it. He was clearly more qualified than I was, with his master's degree in aeronautical engineering. I decided not to mention that I had applied too, because I figured I wouldn't get an interview anyway, and I didn't want my not getting called to take attention away from his accomplishment.

"I do have a favor to ask you," Mark said. "Do you have a suit I could borrow?"

I did—I had just bought a suit to attend a friend's wedding—so I loaned it to him.

Months later, I came back to the office after flying a test flight and my secretary flagged me down. "Hey, Scott," she said excitedly. "You missed a phone call from Teresa Gomez at NASA." Teresa was the long-serving administrative assistant at the astronaut selection office. Her name was widely known throughout the flight-test community; if you got a call from her during the interview process, it was probably good news.

I called back right away, and Teresa asked whether I wanted to come down for an interview. "Yes! Of course," I answered, trying not to shout. "I can come whenever you want."

My interview was scheduled for a couple of weeks later. In the meantime, Mark came back from his own interview feeling that he had done well. He filled me in on exactly what to expect, which was hugely helpful. With Mark's information, I could think about how I would deal with each stage of the daunting process and what answers I might give in the interview. Along with Dave Brown, a fellow Navy pilot who had also interviewed with NASA in the first group, we set up a camera in a conference room at the test pilot school in the evenings to videotape our sessions. Mark and Dave asked me the same questions the committee had asked them, I gave my answers, then we critiqued the video together. "Lean forward more," Dave urged me. "Be more animated." It not only helped me enormously, but it was a pretty nice thing for them to do, considering that we were in competition.

I reminded Mark that the astronaut selection board had already seen the only suit I owned.

"You have to buy me a new suit," I told him. "It's bad enough that we look exactly the same. We can't wear the exact same thing too. They won't be able to remember who said what and we'll look funny."

Mark, being the cheap young Navy lieutenant he was, refused to lay out the money for a new suit. I packed the same suit for my interview.

I was looking forward to heading down to Houston when the government suddenly shut down. This was in the fall of 1995, when President Clinton and the Republican Congress were in a standoff that left the government without a budget, causing shutdowns on and off from November until the following January. NASA was one of the many government agencies that had to temporarily close its doors.

During a window between shutdowns in December, I finally went for my interview. I checked into the hotel near the Johnson Space Center where all the applicants in my group would be staying, the Kings Inn, the same hotel that earlier astronauts stayed in while going through this same process. The interviews and tests would last an entire week, so the interview groups of twenty people got to know one another pretty well. Astronaut hopefuls have an acronym at NASA, just like everything else: ASHOs, pronounced "ass-hoes," but as if an L had been conveniently placed between the O and the E. After checking in, I found some of the others milling around in the lobby and introduced myself to them. The candidates I didn't meet that afternoon I met in the hotel bar that night. There was definitely a sense of sizing one another up as competition. At the same time, we knew we might also be future colleagues and spaceflight crewmates.

The interview and selection process is grueling—intentionally so, I think. We were interviewed, we took written tests, and we went through extensive medical testing. We had even more thorough eye exams than in the Navy, though in this case there was only one doctor in the room, not a team of four, and that doctor made no effort to try to intimidate us.

A lot of the medical tests were ones you would expect to get in a normal physical—blood tests, urine tests, reflex tests, questions about our family histories, that sort of thing. Some of the tests went into more depth than we had experienced before. We had all known to expect this. Obviously, astronauts have to be in exceptionally good physical condition and to have as low a risk of developing health problems as possible. Minor issues could disqualify astronaut hopefuls—for example, one occurrence of kidney stones could disqualify you from flying in space. NASA can't risk a recurrence that would incapacitate an astronaut or require a costly early return. Anyone who had ever slipped a disc, had a heart murmur, or had been diagnosed with any of a number of otherwise generally inconsequential illnesses or injuries was possibly ineligible. Interestingly, a history of gallstones would disqualify an applicant, but not having a gallbladder was fine.

The medical tests could be anxiety inducing. There was nothing we could do to prepare for them or to maximize our chances, other than getting into the best physical condition we could beforehand. I had been running every day at lunchtime since I got the call for the interview. Since the government shutdown had stretched on for months, I ran so much that my heart had started skipping a beat every minute or so. I talked to Dave Brown about it in confidence. He speculated that my resting heart rate was getting so low that the backup mechanism for making sure my heart didn't stop altogether was kicking in. Then my heart compensated by skipping the next beat, resulting in something called a premature atrial contraction (PAC). If that was what it was, it wouldn't pose a threat to my health but might still be enough to disqualify me. With so many applicants, NASA can afford to set aside candidates with even the tiniest chance of developing a health problem.

As part of the process we went through in Houston, we each had to wear a Holter monitor, a device that records the heart's activity, for twenty-four hours. While I was wearing it, I was aware of every time my heart skipped a beat, wondering if it was going to ruin my chances of becoming an astronaut. The NASA flight surgeon assigned to me for the interview process was Smith Johnston. He communicated with me as much as he could without breaking any rules (he wasn't allowed to tell astronaut hopefuls whether or not we were medically qualified). Smith let me know that while my PACs could be an issue, he would do his best to convince the medical review board not to let them hold me back. He also mentioned that my cholesterol was unusually low—who knew cholesterol could be too low?—which I attributed to the rabbit diet I had been on for the last few months. As with the running, I had been so determined not to leave anything to chance that I had almost overdone it.

The week's most memorable test was the proctosigmoidoscopy. It's like a colonoscopy that doesn't go up as far, and without any sedatives or anesthetics. It's painful and humbling, and as with so many other things we went through that week, we wondered how much we were being tested for toughness as well as for medical issues. I remember lying on my side on the examination table when the gastroenterologist came in and greeted me; I noticed there was a TV screen behind him, and on the screen was the image of a pair of shoes. It took me a moment to figure out that I was looking at the tops of the doctor's shoes, and that the monitor was showing the view from the camera at the end of a long, flexible scope in his hand. A split second later, the view changed: now I was looking at my own asshole. It was not a view I had ever seen before (and hope never to see again), and I didn't have much time to contemplate it before the image became an interior view.

Besides being incredibly painful, what made this procedure more unpleasant was that the doctor needed to pump air into me in order to be able to see, and at the end of the procedure when I was allowed to get up and get dressed, that air remained. I was scheduled for a tour of Space Center Houston right afterward, so I walked over there trying not to expel all that air (and other matter) in an attention-getting way. As with everything else, I wondered whether the challenge not to shit my pants in public was part of the test, to see how we would deal with this kind of discomfort and embarrassment. It's true that life as an astronaut, especially on the space station, has more than its share of physical humiliations.

Finally, it was time for my interview with the selection board. I stood in the hallway outside a conference room as Duane Ross, who ran the astronaut selection office, was inside reading an essay I had been asked to write about why I wanted to be an astronaut. As I waited, I remembered the paragraphs I had written and rewritten as if my life depended on them.

> The main reason I want to be an astronaut is that it is the most challenging and exciting job I can imagine. I want to play an integral part of humankind's boldest endeavor ever, and truly feel that I would be an asset to the human space program.
>
> In today's society, our children are in desperate need of role models to inspire and motivate them to excel in sciences and math. The inspiration to explore and achieve the human space program provides to our children today will result in countless intangible benefits for future generations. I want to be a part of this future and feel the human space program would provide the best forum to serve as a role model for our children.
>
> America has always had lofty goals to inspire achievement in all aspects of our lives. In this century we have used human flight, in our quest to fly faster and farther than anyone has ever done before, as a benchmark for technological achievement. The Wright *Flyer,* the *Spirit of St. Louis,* and the Bell X-1 are all examples of great achievements that have inspired previous generations. The human space program is now and forever will be this country's inspiration, and I want to play an integral part in it.
>
> The entire world needs spaceflight to advance scientific dis-

> coveries in medicine, engineering, science, and technology. Just as the Apollo program resulted in countless tangible benefits that improved the daily lives of all individuals, the human space program is necessary if we are to continue our great history of technological achievement. It would be an honor to be a part of any discovery made as a result of the human space program.

I wondered whether I had used the phrase "human space program" too many times. I wanted to show that I understood that human spaceflight is not the only thing that NASA does, and also that I knew the phrase "manned spaceflight" was outdated. I had found this essay fiendishly difficult to write, because I knew my answer to the question "Why do you want to be an astronaut?" would be more or less the same for everyone else. We all wanted to do something difficult and exciting and important. We all wanted to be involved with something that would be in the history books for hundreds of years to come. What more was there to say about it? How could one applicant differentiate himself or herself from the others? Now that I have served on the astronaut selection board, I know that the essay doesn't do much to help or hurt an applicant unless it is extreme in some way. But at the time, every detail seemed important.

In an earlier draft, I had tried, just as an experiment, to be more honest and see how it sounded.

"Actually the real reason I want to be an astronaut is that when I was in the tenth grade and visiting Kennedy Space Center on a family trip I wanted to see the film about the manned space program. My parents said the line was so long we would only go if Mark and I were in it."

I had looked the new paragraph over with a stern eye, then decided it was too big a risk to try to be funny or cute. I stuck with the original message. It might be a cliché, but it was true.

I knew from Mark and Dave that an intimidating group of twenty people would be interviewing me. Some of them I recognized. John Young was one of them, the only astronaut to have launched on three different spacecraft: Gemini, Apollo, and the space shuttle. He had orbited the moon alone on Apollo 10, then walked on its surface dur-

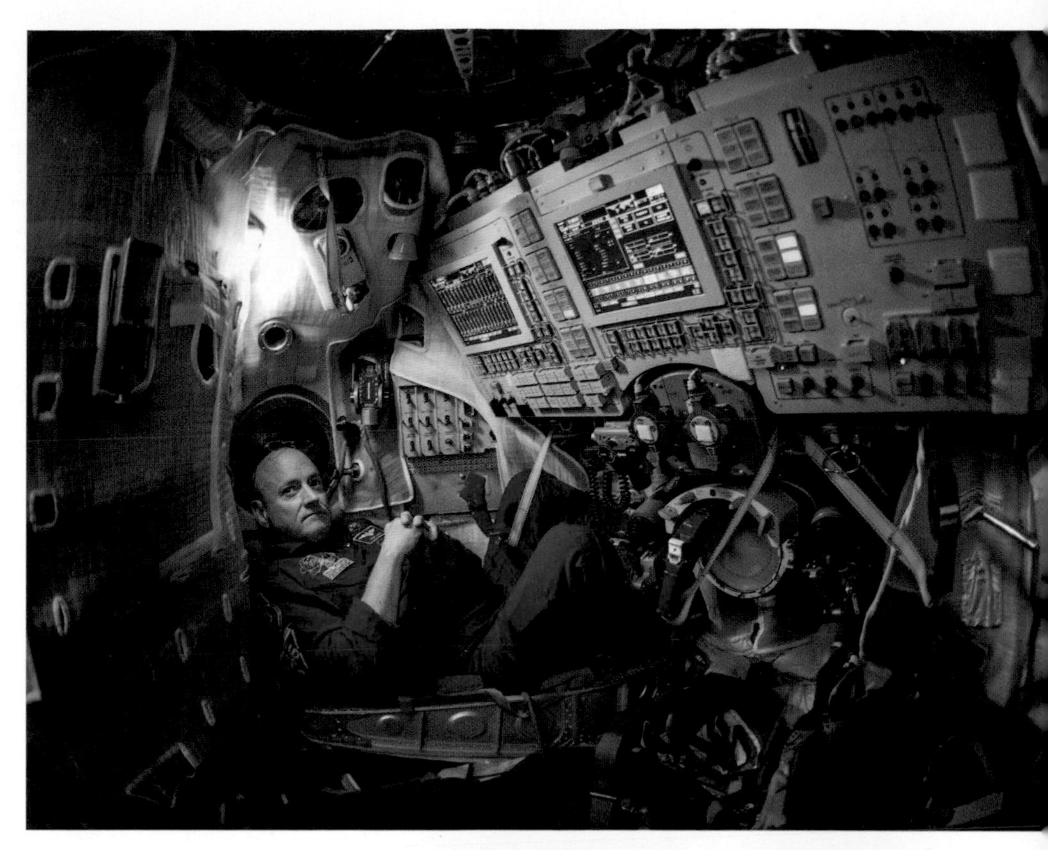

Reflecting on my pending yearlong space mission from inside a Soyuz simulator at the Gagarin Cosmonaut Training Center outside of Moscow

With my Russian cosmonaut crewmates Mikhail (Misha) Kornienko, center, and Gennady Padalka, being interviewed by the media in front of the Lenin statue in Star City, a few days before our departure for Baikonur, Kazakhstan, site of the launch

With Gennady, center, and Misha a few days prior to launch. In the background, the "high desert" of Kazakhstan

Our Soyuz TMA-16M spacecraft rolls out by train to the historic launchpad at the Baikonur Cosmodrome.

Russian Orthodox priest Father Job blesses the Soyuz rocket that will take us to the ISS.

Gennady is on my left, and Misha is on his left, as we share our last traditional Russian breakfast with our backup crew before the flight.

I chat with Gennady as we wait for final pressure checks of our Sokol suits.

My crewmates and I talk to the media as our families look on from behind a glass window while we're in quarantine prior to heading to the launchpad.

On our way to board a bus that will take us to the launchpad

We wave good-bye to friends and planet Earth as we board our Soyuz spacecraft before launch.

Our Soyuz spacecraft lifts off toward space and the ISS on March 28, 2015.

While flying about 250 miles above Earth, we used our station's robotic arm to capture SpaceX's Dragon cargo craft filled with experiments and supplies for our crew.

On the ISS with my crewmates Anton Shkaplerov, left, Samantha Cristoforetti, center, and Terry Virts, celebrating Samantha's birthday

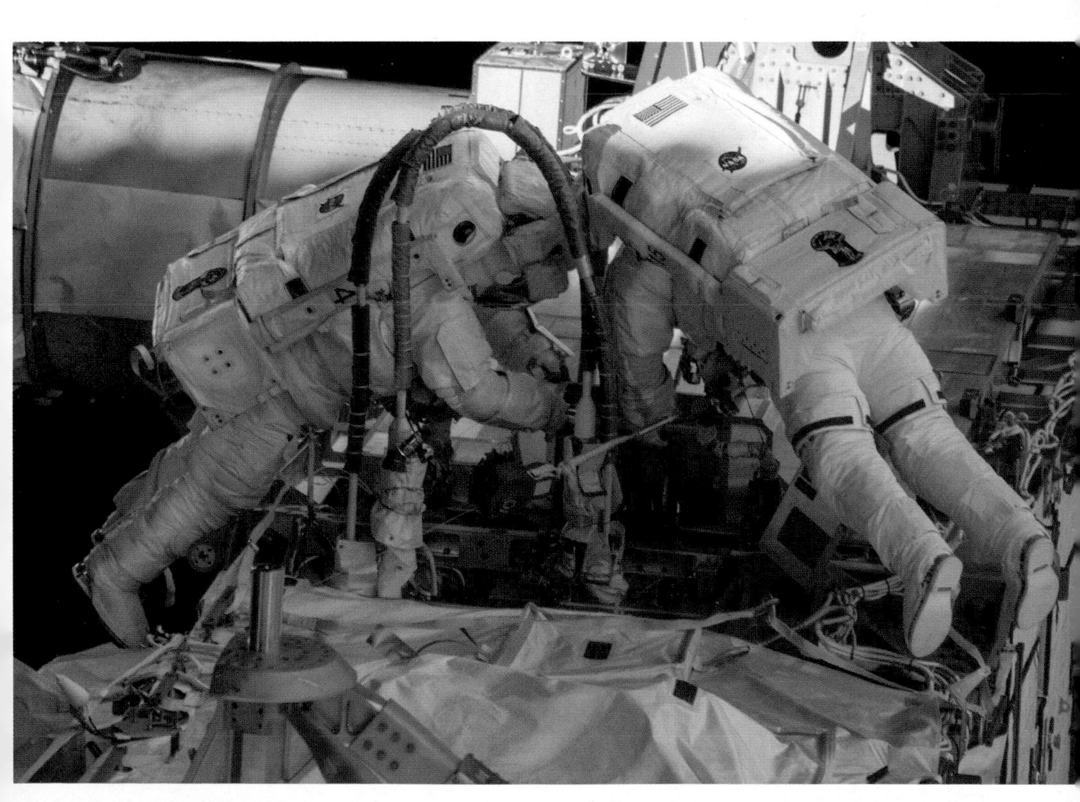

With Kjell Lindgren, right, as we work outside the ISS to restore the port truss ammonia cooling system to its original configuration on our second spacewalk, which lasted seven hours and forty-eight minutes.

My feet get in the way as I admire the unmistakable blue waters of the Bahamas from the Cupola aboard the ISS.

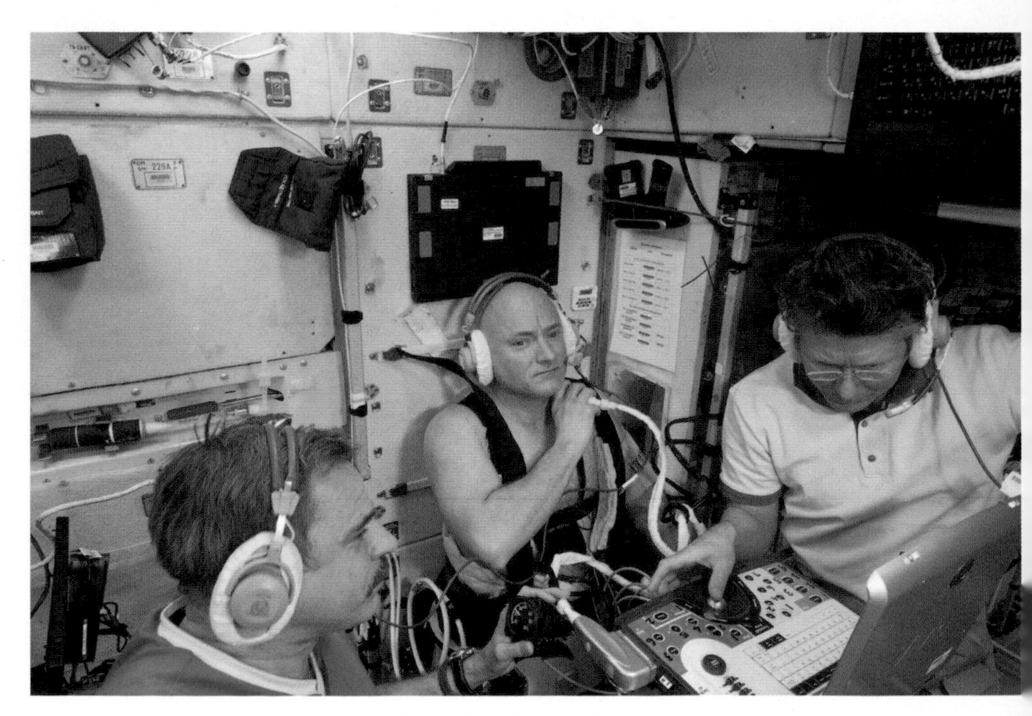

Misha and Gennady help me perform the Fluid Shifts experiment in the Russian service module.

ing Apollo 16. He was chosen to command the first flight of the space shuttle, making him and his pilot, Bob Crippen, the only two people to launch into space on a rocket that hadn't been previously tested on unmanned flights. He was what you might call an astronaut's astronaut, a living legend. I wanted to be just like him. I also recognized Bob Cabana, the chief of the Astronaut Office (he had greeted us a few days earlier), and astronauts Jim Wetherbee and Ellen Baker.

I got settled into a chair at a T-shaped table surrounded by the committee and tried to sound calm and confident as I greeted them.

"I'm afraid this might all look pretty familiar to you guys," I said, pausing for a laugh. "You've seen this suit before." Then I explained how I had loaned it to my brother, who had been too cheap to buy me a new one. But he did lend me his shoes.

It's risky to try to make a joke in a job interview, but everyone laughed, which made me feel a bit more at ease. They might have been wondering how Mark and I would deal with being twins applying at the same time, and I wanted them to feel they could treat us like any other candidates.

John Young took the lead. He said, simply, "Tell us about your life."

My mind raced. What aspects of my life did he want to hear about? How far back should I go?

"Well, when I graduated from college in 1987—," I began.

"No," Young interrupted. "Go back further. Go back to junior high school."

In retrospect, I wonder whether they cut off everyone's response and made them start in a different place, to see how they would respond to being interrupted. In my case, junior high was not a great place to start. I wasn't about to tell them about staring out the window and earning C's. So instead, I told them about fixing up boats with my father, about learning to be an EMT and the experiences I had had working on the ambulance, about becoming licensed as a Merchant Marine officer in college, about learning how to work in an operational environment, and about the challenges I faced along the way. As I spoke, I was trying to put my experiences in a context that would differentiate me from the other candidates they were seeing. Being a test pilot, as tough as that had been to achieve, wasn't going to set me apart from the other test pilots. But repairing a clunker boat in the open waters of the Atlantic might, or delivering a baby in a roach-infested slum in Jersey City.

"What's the frequency response of the longitudinal flight control system in the Tomcat?" John Young asked.

I had mentioned in my application that I was working on a new digital flight control system in the F-14. I had also been tipped off by an F-16 pilot who interviewed the same week as my brother that Captain Young liked to ask about longitudinal frequency response, so I was prepared.

"Fifty hertz," I answered.

Young nodded approvingly. It would make sense that the selection board would want to see whether or not I knew my stuff technically, but I also think he was just fascinated by planes and never stopped wanting to learn about them.

The official interview lasted around fifty minutes, with Captain Young and Bob Cabana leading the conversation. In general, I felt that I was doing well, though at one point I noticed that Ellen Baker looked like she might be falling asleep. My interview was the first one after lunch, and I hoped that it was something she had eaten that was making her eyes droop and not my boring stories.

PART OF the selection process involved psychological tests, which I found interesting but stressful, since so much was riding on them. I was tempted to try to figure out what the "right" answer for each question was. The answer to "Do you ever hear voices telling you to do things?" wasn't hard to guess, but I figured the test was designed to reveal people choosing to lie. One question I remember specifically was "Would you rather steal something from a store or kick a dog?" I had to choose one, so I said I would rather steal something. With that type of question, I suspected there was no right or wrong answer, but rather that our responses would be cross-checked against other responses to similarly worded questions to detect someone trying to game the test. Years later, one of the psychiatrists told me I'd almost failed the test for that reason—my answers reflected that I was trying to tell them what they wanted to hear.

Other tests were more unusual. Because an astronaut cannot be claustrophobic, we all underwent a simple test: we were each outfitted with a heart monitor, zipped into a thick rubber bag not much bigger than a curled-up adult, shut into a closet, and left without any idea of how long we would be in there. For me it was about twenty minutes, and I enjoyed a brief nap. Another requirement was that we go over to the Astronaut Office at some point during the week and chat with some of the astronauts. I dutifully went over one afternoon, introduced myself to the first person I met, and then got out of there fast. I figured there was not much upside to this exercise: in a brief conversation I wouldn't be able to make a good enough impression to help me, but I could easily rub someone the wrong way and wind up hurting my chances.

One of the events on our schedule was a dinner at Pe-Te's Barbecue, a popular destination for off-duty astronauts and other NASA employees. This dinner was one of the more informal events, and in some ways that only made it more stressful. I figured the selection board didn't want someone who was sloppy or unpresentable when off the clock, but I also didn't want to look like an uptight square who didn't know how to have a good time. I thought longer and harder about what to wear that night than I did for any other part of the selection process. I actually looked up photos of astronauts at casual events to see what they wore. Based on my investigation, I chose khaki pants and a Ralph Lauren striped polo shirt. At the restaurant, I was faced with more daunting choices: Should I drink only water, to show how health-conscious I was? Should I drink one beer to show I could stop at one, or drink two to show I could stop at two? The social terrain was tricky as well. Should I talk to astronauts as equals and risk sounding disrespectful, or treat them as superiors and risk sounding like a suck-up? Should I avoid talking to them and risk having them not remember me at all? I could see all the other candidates around me making the same calculations.

AT THE END of the week, we said our good-byes and went back to our respective homes. NASA was to interview six groups in all, and I had been in the third group, so patience would be required. For me, the wait was made harder by the fact that I thought I had done well. If I'd known that I'd blown some part of the process, or that one of the doctors I'd encountered had betrayed that something was wrong with me physically, I would have had a pretty good idea that I wasn't going to make it, and the wait would have been easier.

As the weeks went by, I received a new Navy assignment: to join a fighter squadron at a naval air station in Japan. This was a move I would have been excited about under any other circumstances, and Leslie was prepared for the adventure, but I still hadn't heard from NASA. I didn't want to move my family there until I had to.

The moving company that contracted with the Navy called me to set up a date to come and pack up our stuff.

"Can you hold off for a couple of weeks?" I asked. The movers reluctantly agreed.

Soon, they called again. This time they had chosen a date they wanted to come and they were less interested in renegotiating it. I managed to put them off again. And again. Over the next few weeks, I started to hear from some of the people I had listed as references that they had been contacted as part of my background investigation. So I knew I had made it to the next level. That gave me hope, though I was still concerned about the fact that I was interviewed in the third group rather than the first. I asked people I had met at the interview if they had heard anything about when NASA would make their calls. No one did.

A few days before Memorial Day weekend, I got a call at home.

"Scott," the voice said. "This is Dave Leestma." Dave was one of the astronauts I had met in Houston and was the flight crew operations director—the direct supervisor of the chief astronaut.

"Yes, sir," I replied.

"Would you like to come fly for us?" he asked. I paused, because it wasn't entirely clear to me that this was the call I had been waiting for. I knew that NASA employed a lot of pilots who weren't astronauts—maybe Dave was asking me to be a pilot, not an astronaut.

"Uh, maybe," I said. "Fly what?"

He answered with a laugh. "The space shuttle, of course."

It's hard to describe what I felt then. I wasn't entirely shocked, because I thought I had done well, and I had started to think I might be chosen. But I did feel an awareness of everything it had taken for me to get here, from reading *The Right Stuff* and setting a goal that seemed impossible, to this moment. And I felt humbled by the role I was going to be asked to step into.

"I'd love to," I said. "Have you called my brother yet?"

Later, when I related this conversation to people, they thought it was funny that I didn't even take a breath to process my own accomplishment before asking about my brother. But to me, waiting to find out what would happen with his application was almost as suspenseful as waiting to hear about my own.

"I just got off the phone with him," Dave answered. "Yeah, he got selected too."

This was the first time NASA had selected relatives. We'd been concerned they might not want to select brothers, especially twins, and in the back of my mind I had been anticipating that they might choose one of us and not the other.

"Mark actually asked me about you, too, and I told him I was about to call you," Dave said. So my brother knew that I was to become an astronaut before I did. That was fine with me.

I hung up the phone after talking to Dave and I told Leslie: "I'm going to be an astronaut." She was thrilled for me. Next I called my brother, and we spent a few minutes on the phone congratulating each other and talking about our moving plans. I got on the phone with my parents and they were overwhelmed by the news. Word spread quickly within our small family—the next time we saw our maternal grandmother, she had had a custom bumper sticker made for her car that read, MY TWIN GRANDSONS ARE ASTRONAUTS. I would imagine people thought she was crazy.

The next day, I told my colleagues that I had been chosen to be an astronaut. I particularly enjoyed telling Paul, my friend and flight test engineer, because I knew he would be excited for me. When I told him, he jumped up and, with a huge smile, exclaimed, "You've gotta be frigging shitting me!" A few seconds later he followed up with "Will you invite me to come down to Florida and see a launch?" I promised I would. I was surprised and touched by how pleased everyone was for me. They were all so thrilled, their excitement actually helped it to sink in for me what I had achieved. My life had just changed. I was going to have the chance to fly in space.

When the press found out that NASA had selected the two of us, they called the astronaut selection office to ask about it. A reporter asked Duane Ross, "Did you know you picked two brothers?"

His answer: "No, we picked two very accomplished test pilots who happen to be twins."

12

ONE HOT DAY in early July 1996, Leslie and I packed up our two cars and left Pax River for Houston. Samantha, almost two now, was a sprightly and adorable toddler. We found a house we liked quickly and moved in on August 1. Mark and his family moved to town after we did, since they were having a house built nearby.

In addition to getting my family settled and learning about the area, I was also working out a lot, running every day. I wanted to show up at NASA in good shape. There was part of me that felt like I was still trying out for the job, and in a sense I was—I hadn't been assigned to a flight yet. I still thought of myself as a below-average guy stepping into an above-average role, and I knew I would have to impress some people if I was going to be among the first in my class to fly.

On the Friday night before our official Monday start date, we went to a party where we met all of my new classmates. We were ASCANs (pronounced "ass cans"), short for astronaut candidates (we would become full-fledged astronauts the first time we left the Earth's atmosphere). The party was hosted by Pat Forrester, who was selected in our class but had already been stationed at NASA as an Army officer. Because he already knew his way around, he was our official class leader.

It wasn't until that party that I learned our class would include international astronauts. There were thirty-five Americans and nine astronauts from other countries, which made us the largest astronaut class in NASA history. At the party, I was chatting with Mark and some other new classmates when I heard a man nearby I hadn't met before, who was speaking with an accent. I figured he might be one of my foreign classmates, so I went up to him, stuck out my hand, and said, "Hi, I'm Scott Kelly."

Before he could answer, a woman pushed him out of the way, stuck her hand out, and said, "*I* am your classmate. My name is Julie Payette." The man she had pushed aside was her husband. They were both French Canadian, bilingual in French and English, and she had grown tired of people assuming her husband was the new ASCAN rather than her. She and I would go on to become great friends. I met so many people that night—not only my classmates, but their spouses and significant others, astronauts from previous classes, *their* partners, and other NASA people who worked in support of the Astronaut Office. It was exciting to know that we were going to be such a big part of one another's lives, and maybe spend time in space together.

The first day on the job involved a lot of paperwork and learning the basic aspects of working for NASA. Jeff Ashby was the astronaut from the previous class in charge of getting us oriented. We were introduced to the rest of the Astronaut Office and shown where our desks would be. I was to share an office with my classmates Pat Forrester, Julie Payette, Peggy Whitson, and Stephanie Wilson.

Our training started out in classrooms, where all forty-four of us began to realize the magnitude of knowledge we were going to need. We heard lectures on geology, meteorology, physics, oceanography, and aerodynamics. We learned about the history of NASA. We learned about the T-38, the jets the astronauts fly.

Most of all, we learned about the space shuttle. We were given an overview of how the shuttle worked as a whole, and we got specific lectures on each of the many individual systems—their designs, their nominal operations, their possible malfunctions, and how we should respond to those. We worked through a number of different failures that could occur as we executed the procedures we would use on actual missions. We trained that way on the main engines, on the electrical systems, on the environmental control and life support system. It was challenging to master all of it, but it became even harder when we moved on to the shuttle mission simulator, which integrated all these systems together during the mission phases: prelaunch, ascent, post-insertion, on-orbit operations, deorbit prep, entry, landing, and post-landing.

Our trainers hammered us with the malfunctions we could face during a real flight. A critical phase was post-insertion, the period of time just as the shuttle is getting into orbit. We have to convert a vehicle that has launched as a rocket into a working orbital spaceship—reconfiguring the computers, getting the enormous payload bay doors open so their radiators could cool the shuttle's electrical systems, deploying the Ku-band antenna so we could communicate with the ground, deploying the robot arm, making sure everything was working properly, and getting ready for on-orbit operations.

By far the most challenging and complicated phase of shuttle training was ascent. On a real launch, when everything went right, the flight crew had very little to do besides monitoring the systems, but NASA had to prepare us for every eventuality. So this phase of flight revealed those who had learned their stuff and those who hadn't. We trained for the orbit phase, since that was where we would spend the most time on a real mission. We practiced payload operations—for instance, deploying and then retrieving a satellite. We practiced rendezvous and docking with *Mir* (the International Space Station didn't exist yet).

We trained to do deorbit prep, which is post-insertion in reverse: learning to take an orbiting spaceship and reconfigure it into something that could reenter the Earth's atmosphere and land—a space plane. We worked on putting the antenna and robot arm away, closing the payload bay doors, getting the computers configured for the last phase of flight, then programming the deorbit burn to slow us down by just a few hundred miles an hour, which is enough to get us to reenter the atmosphere. As a pilot, I practiced reentry and landing thousands of times. We never stopped practicing. This is the moment in the mission when having something go wrong can be the most serious, so I had to be prepared to deal with anything. I remember the first reentry simulation I ever did: I was sitting in the pilot's seat and an experienced astronaut was monitoring me. I felt a lot of pressure to perform well, since this was my first time trying to demonstrate my fledgling astronaut skills in front of a real astronaut. I messed up starting the auxiliary power units, which provided power for controlling the shuttle's three engines and for moving the control surfaces on the shuttle, like the aileron, rudder, and flaps. The APUs lowered the landing gear and powered the brakes, so we couldn't land without at least one. Because of the way I started them up, one of them probably would have exploded. Not a great start. I didn't do particularly well at following procedures verbatim, either. I had been under the impression that the detailed procedures we were learning were more like guidelines; I was wrong. To top it all off, my landing was bad enough that it might have killed us all. The space shuttle is one of the hardest planes to land ever, so on that I got a bit of leeway. On all the other screwups, not so much.

The very complexity of the space shuttle was why I wanted to fly it. But learning these systems and practicing in the simulators—learning how to respond to the myriad of interrelated malfunctions in the right way—showed me how much more complicated this spacecraft was than anything I could have imagined. There were more than two thousand switches and circuit breakers in the cockpit, more than a million parts, and almost as many ways for me to screw up.

The amount I learned in order to go from a new ASCAN to a pilot on my first mission was, from what I could observe, an education comparable to getting a PhD. Our days were packed with classes, simulations, and other training. In the evenings, I would have a quick dinner with Leslie and Samantha, then get back to work studying. I went over notes from lectures and made a training notebook for myself that I could continue to study and add to as my education progressed. I spent at least one full day each weekend going over all of this material.

We went on field trips to different NASA centers—Ames in California, Glenn in Ohio, Goddard in Maryland, Michoud in Louisiana, Marshall in Alabama, headquarters in D.C., Kennedy in Florida. We needed to learn about what happens at each of these sites and how all of NASA's projects work together, even the ones that didn't directly affect the shuttle. As astronauts, we were going to serve as the public face for NASA, and we needed to be able to talk about everything NASA does.

At the same time, it was important that the workers at these sites knew us as human beings whose lives would depend on their work.

My class had earned a reputation by this point for asking a lot of technical questions whenever we got the chance. In an atmosphere where forty-four people are vying for a small number of flight assignments, one of the ways to make an impression on our management was to ask complex questions that made clear how hard we'd been studying and what a strong grasp we had on the technical issues. Just before we went to Ames, NASA's center for aerodynamic research, we were in a lecture when C. J. Sturckow, an astronaut from the previous class and a Marine Corps officer, burst into the room wearing his Marine camouflage uniform.

"Listen up," he said from the front of the room. He took a giant knife out of its sheath and slammed it down on the table. "Everyone is getting tired of all of your questions! You think you sound smart, but you're just slowing things down. When you go to Ames in a few days, I only want to hear yes-or-no questions like 'Is this the biggest wind tunnel you have here at Ames?'" With that, he picked up his knife and left the room without uttering another word. Some people in our class were offended or weirded out by his militaristic display, but I appreciated the directness.

Generally speaking, each of us would be actively training for a mission every few years. In between, we had specific responsibilities within the Astronaut Office. Most of us were put in charge of a system on the shuttle: we were to learn everything about that specific system, take part in redesigning it or improving it, and represent the astronaut's point of view with the engineers. This practice has been ongoing since Gemini days, when the spacecraft first became so complicated that it was impossible for one astronaut to know everything.

I was put in charge of the caution and warning system on the space station, which sounds pretty important until you consider the fact that the space station didn't yet exist. I was trying to learn as much as I could about the space shuttle, because it was the vehicle I was preparing to fly. For the pilot and commander, there are so many seemingly insignificant errors that could result in the loss of the vehicle and crew—it was the most important thing for me to learn not to make those mistakes. So the space station was going to take a backseat in my mind.

Some of us were also assigned to phases of flight to gain a special expertise—in my case, the rendezvous phase. I was pleased with that, because I knew there was a good chance I would fly a mission that would rendezvous with a space station or a satellite someday, and this way I would be well prepared. I would receive rendezvous training well ahead of my classmates, which would have ramifications going forward.

The Astronaut Office was a busy place in those days with such a large new class adding our numbers to the already existing corps. Some very experienced astronauts were still around, and it was an honor to serve along with them. John Young, the Gemini-era astronaut who had been on my selection committee, was always in the astronaut gym, putting everyone to shame just by showing up. Another spaceflight legend, John Glenn, was assigned to his space shuttle flight not long after I became an astronaut. One day I had four-year-old Samantha with me at work because Leslie had a dental appointment, and as I was walking her around I saw Glenn working diligently in his office. I introduced myself and Samantha.

He looked up and said, "Hi, young lady. What are you doing today?"

"I'm going to lunch with my dad," Samantha answered.

"What's your favorite food?" he asked her.

"Macaroni and cheese," she said.

Senator Glenn gave her a look of pleased surprise. He held up the papers he had been working on.

"Look right here," he said. "I was just selecting my space food for my mission and I just wrote, 'Macaroni and cheese.' That's my favorite, too!"

Another time, I had Samantha with me at a party, and I encouraged her to talk to John Young about his experience walking on the moon. Samantha approached him and said, "My dad says you walked on the moon."

John responded, "I didn't *walk* on the moon. I *worked* on the moon!"

More than a year later, we were watching a documentary about

Apollo, and I pointed out John Young to Samantha. "You met him, remember? He walked on the moon."

Samantha didn't miss a beat: "Daddy, he didn't walk on the moon, he *worked* on the moon."

John Glenn completed his mission, in October 1998, after which I inherited his parking space and used it for the next eighteen years.

Leslie and Samantha took easily to life in Houston. Leslie was always good at making new friends, and she quickly became part of a tight circle of women in our neighborhood. I would often come home from work to find a group of five or six women clustered in the kitchen, drinking wine and eating cheese, talking and laughing. She also became the head of the astronaut spouses' group, which was responsible for planning social events for the astronaut corps, especially the traditional parties in honor of the spouses of the crew that was flying next. They also helped out with meals, babysitting, and other favors for anyone in the group who had a special need, like a death in the family or a new baby. The role suited Leslie well.

As PART OF my ASCAN training, I learned to fly the Shuttle Training Aircraft (STA), a Gulfstream business jet that had been modified to recreate the approach profile and handling qualities of the space shuttle in the landing phase as closely as possible. Flight computers simulated the drag we would experience in the heavier, less agile orbiter by putting the engines in reverse while airborne. The left side cockpit and the controls had been designed to simulate the experience of landing the shuttle. The STA generally flew out of El Paso, Texas, so we would fly over there in a T-38, which took a little over an hour, get in the STA, then fly another thirty minutes to the White Sands Test Facility in New Mexico. I did many practice approaches to the dry lake bed runways in that aircraft, stopping short of actually touching the wheels down. At first, we'd fly the STA every few weeks, learning to land the space shuttle. Eventually, we moved to flying every other month, then every quarter to maintain our proficiency, until we were assigned to a real mission.

I was in El Paso one day in March 1999, just having finished my ten practice landings and getting ready to fly back to Houston, when one of the senior shuttle commanders, Curt Brown, a tall guy with a receding hairline and a thick Tom Selleck 1980s mustache, came up to me. He had only spoken to me a couple of times before. He was known to be extremely technically competent, and his experience—five shuttle missions in six years—was nearly unequaled. But he also had the reputation of being arrogant and unfriendly to those not worthy of his attention. A high flight rate, training for missions one after another, practically without a break, can also bring burnout.

"Hey, come over here," he said sternly. "I need to talk to you."

I followed him into a private office, wondering what I had done to piss him off. He shut the door behind us, then turned and poked me in the chest three times while staring straight into my eyes.

"You better have your shit together," he said to me, "because we're flying in space in six months."

I felt a couple of different things at once. One was: *I'm fucking flying in space in six months!*

Another was: *Wow, what a shitty way to let someone know he's got his first flight assignment.*

"Yes, sir," I said. "I've got my shit together."

Curt told me to keep this news a secret. I told my brother, of course.

A couple of days later, I was called in to see Charlie Precourt, the new chief of the Astronaut Office, along with Curt and French astronaut Jean-François Clervoy (we called him "Billy Bob" since "Jean-François" didn't sound very Texan). Charlie looked very serious. He told Billy Bob and me that we were in trouble. A few months earlier, he said, we had screwed up on a T-38 flight and had drawn a flight violation from the FAA.

Because of my run-in with Curt earlier that week, and knowing that we were being assigned to a flight, I had a strong feeling that he and Charlie were just messing with us. Billy Bob didn't know that, though, and all the color drained from his face. Once Curt and Charlie had had enough fun, Charlie said, "We're just kidding, guys. You've both been assigned to STS-103 on *Discovery.* It's going to be an emergency repair mission to the Hubble Space Telescope."

Billy Bob was visibly relieved. Curt would command the mission, and we would be joined by John Grunsfeld, Mike Foale, Steve Smith, and Claude Nicollier. I was to be the only rookie on the crew and the first American in my class to fly. The primary goal was to fix the failing gyroscopes on the Hubble Space Telescope on four spacewalks, each more than eight hours long. Hubble needs at least three of its six gyroscopes to be working in order to make precise observations, and three had already failed.

The Hubble Space Telescope has been making observations of the universe since 1990. Until then, astronomers could never get a truly clear view of the night sky because of the distorting effect of the atmosphere, the same effect that causes stars to appear to twinkle. Observing stars and galaxies through the filter of our atmosphere was like trying to read a book underwater. Putting a telescope in orbit outside the atmosphere and past the reach of light pollution has changed the field of astronomy. By observing distant stars, scientists have been able to make discoveries about how fast the universe is expanding, how old it is, and what it is made of. Hubble has helped us to discover new planets in new solar systems and confirmed the existence of dark energy and dark matter. This one scientific instrument has revolutionized what we know about our universe, and the task of repairing it—which always brings the risk of damaging or even destroying its sensitive components—is an enormous responsibility.

Once our training was in full swing, we spent a lot of time in simulators. Running simulated missions is the only way for astronauts to get hundreds of hours of experience doing something that in reality we would get to do only a few times. The simulations re-created the experience as closely as possible—same screens, switches, and buttons; same uncomfortable metal-framed seats, same headsets, and same thick procedure books. The simulation supervisors devised fiendish scenarios for us to work through, such as multiple interrelated systems failing while other systems continued working just fine, though their sensors might erroneously report they had failed too. We practiced solving problems quickly. Often the simulations were designed so that one of us would be barraged with problems to test how we worked together as a team.

About halfway through our training, we were in a simulator dealing with a complex failure—all the cooling systems had gone down at once. Those controls were all on the left side of the cockpit, where the commander, Curt, was sitting. He was hit with one malfunction after another, but because he was so talented and experienced, he was able to identify and focus on the most critical issue. Simultaneously, a computer failed. This would normally be his responsibility too, but because I wasn't as busy and could reach his keyboard myself, I decided to fix it for him by switching out the backup for the primary system. I typed in the commands while Curt's head was still buried in cooling system problems. *Item 16, execute,* I typed.

A few minutes later, Curt got through his work with the cooling system. He looked at the display and saw that the computer failure had disappeared. He looked confused.

"What happened to the port failure on FF One?"

"Oh, I port-moded it for you," I answered. As I spoke, I sensed this was not the answer he wanted to hear.

"You did *what*?"

"I port-moded it."

A second went by—and then Curt turned toward me, which was difficult to do wearing a pressure suit while strapped tightly into his seat. He punched me on the arm as hard as he could.

"Don't ever do that again!" he shouted.

"Ah, okay," I said. "I won't ever do that again."

He'd made his point, and though I didn't agree with his method, I appreciated his directness. I never touched any buttons or switches on his side of the cockpit again without his explicit approval.

EILEEN COLLINS BECAME the first woman to command a space shuttle mission, on *Columbia,* in July 1999. Once that flight got off the ground, we would become the prime crew, our launch date set for

October 14, 1999. But there was a problem on *Columbia* during ascent. An electrical short disabled the center engine's digital control unit. The engine continued to operate on its backup—a case of NASA's redundancy saving the crew from what would have been a very risky attempt at an abort—but something had gone seriously wrong, and NASA needed to find out what it was before flying again. The *Columbia* mission was cut short, and when the shuttle was safely back on the ground, an investigation ensued.

It was revealed that wiring in the payload bay had been chafing against an exposed screw, a good reminder to everyone of how little it can take to cause a disaster. Further inspections revealed deteriorating wiring throughout the space shuttle fleet that would need to be addressed before any of the shuttles could fly again. That caused a delay in our launch date to November 19. As inspections and repairs to the wiring dragged on, we were delayed further, to December 2, then to December 6.

These delays were frustrating for everyone. It was mentally draining to keep working toward a date that slipped away, then bring our full energy to the next announced date. The December 6 launch date didn't change as November went by, and we grew hopeful. We celebrated Thanksgiving with our families, then the next day we said our goodbyes and went into quarantine. NASA's quarantines were a bit different from the Russians'—they were more stringent in some ways and less stringent in others—but the underlying concept was the same: to isolate space travelers from germs before a launch in order to decrease the chances of us getting sick in space.

There were crew quarters at both Houston and the Cape, very similar in style to each other, where quarantined astronauts live. In both places, the crew quarters were more like an office than a hotel—spartan accommodations. The time when the shuttle was to rendezvous with the telescope would be in the middle of the night Florida time, so we had to adjust our sleep schedule significantly. In order to help us make the adjustment, the crew quarters had few windows, and the lights were kept glaringly bright during our waking hours. There were cooks to make us food and a gym to work out in.

We didn't have a great deal to do once we were in quarantine—we had our checklists (about five feet tall when stacked on one another) to review. We had some of the spacewalking hardware and photography equipment to familiarize ourselves with. We had to sign crew photos to hand out to people who worked on the mission, at least a thousand of them. At the end of our workday, which was actually in the morning, we watched movies together.

While we were in quarantine, our launch date changed again, from December 6 to December 11. It was mildly annoying to know I had spent four days in quarantine that I could have spent at home, but we all understood that delays were part of spaceflight. Then we were delayed again, to December 16. By the morning of the sixteenth, we had been in quarantine for twenty days and were getting tired of it. We were ready to go to space or go home. Then the launch was scrubbed. Inspectors had found a possible problem with a weld in the external tank. Workers needed a day to make sure the issue had been resolved, so we were delayed to December 17.

That morning, I woke up and looked at the weather forecast. There would be a low cloud ceiling, rain, and possibly even lightning. The prediction of weather favorable for launch was only 20 percent go, not very good odds, but the weather in Central Florida could change quickly, so the countdown continued. Workers began filling the external tank, a process that takes hours. We got suited up and headed out to the launchpad. The countdown still continued; it seemed we might finally be going to space. We got strapped into our seats and started preparing the space shuttle to launch, the countdown continuing toward our planned liftoff time of 8:47 p.m. There are a few "holds" built into the countdown—points where extra time has been allowed so we can stop the clock and make sure everything is being done right without being rushed. One of these holds is at T-minus nine minutes, and it's the last chance to review all the factors that go into deciding whether we are "go" or not. We kept at the T-minus nine hold for a long time, up to our planned launch time and past it. At 8:52 p.m., the launch director made the decision to scrub due to weather. We would try again the next day.

On December 18, we scrubbed again, this time without suiting up.

At this point we had been in quarantine twenty-two days. If we'd known from the start how many delays were in our future, we would have gone back to Houston to do some refresher training in the simulators and see our families. Because I was launching for the first time, I had invited practically everyone I knew to come to Florida, along with their friends, about eight hundred people in all, and with every delay the group got smaller as people changed their travel plans. The morning of each launch attempt, friends and family would call and ask, "What are the odds you're going to launch today?" I understood their impatience, but I never knew what to tell them. Eventually, I started to just say, "Fifty-fifty. Either we'll launch today or we won't."

Jim Wetherbee, an astronaut who was serving as the director of flight crew operations, came by to talk to us. We all sat around a conference table together, and Jim said, "We're going to knock this thing off and try again in the new year." It was now a week before Christmas, and NASA had decided to give the ground crew a chance to go home to their families for the holidays. We were also coming up against another type of conflict: NASA wanted us safely back on Earth before January 1, 2000, because there was so much anxiety about whether equipment would continue to work properly because of Y2K. We joked that NASA was concerned the space shuttle computers would divide by zero and we would travel through a wormhole and end up on the other side of the universe. But the truth was less exciting. Specifically, the concern had to do with the possibility that we would have to land at Edwards Air Force Base in California. The ground support equipment at Kennedy Space Center was all Y2K compliant, as was the orbiter itself, but the equipment at Edwards had not yet been certified. Personally, I thought the public would find it reassuring if NASA, the agency that had put a man on the moon and created a reusable space plane, was so little concerned about Y2K that they flew in space anyway.

"We haven't made a definite decision yet," Jim said. "But we're ninety-nine percent sure this is what we're going to do."

He left, and we talked about what this delay would mean for each of us. All of my crewmates seemed pleased—they wanted to go home. I was the only one who didn't want to see the launch postponed. I had come here with the expectation of going to space, and I didn't want to give that up and wait weeks before we actually launched. We packed up our things. The guy who holds our wallets for us while we fly in space came around to hand them back out, which made the decision feel final. I prepared to head back to Houston.

Jim came back about an hour later and gathered us together. "Okay, guys," he said. "We changed our minds. We're going to launch tomorrow."

This was tough on my crewmates, who had mentally checked out and started looking forward to going home. I was the only one who was happy, because I was the only one who had never been to space before.

The next day, December 19, as promised, we got suited up for launch. The weather was only 60 percent go, but the countdown continued throughout the day. Several hours before the scheduled launch time of 7:50 p.m., we left the Operations and Checkout Building and waved to the media as we walked to the Astrovan, an Airstream motor home that is used exclusively for carrying astronauts the nine miles to the launch site. The space shuttle, fully loaded with liquid oxygen and hydrogen, was essentially a giant bomb, so when it was fueled the area was cleared of nonessential personnel. As we approached the launchpad, which usually bustled with hundreds of workers, we saw that it was eerily abandoned, the emptiness juxtaposed with the noise of a fully fueled space shuttle—pumps and motors spinning and the creaking metals reacting to supercooled propellants.

We rode the elevator in the launch tower up to the 195-foot level, and Curt entered the orbiter first. The cryogenic fuel passing through the propellant lines created condensation that froze into snow, so even though the weather was warm, some of us had a brief snowball fight while others used the bathroom known affectionately as the Last Toilet on Earth.

Then we entered the White Room one by one, a sterile space around the hatch. When it was my turn, I got into the harnesses for my parachute and fitted the comm cap on my head. Then I kneeled just inside the hatchway while the closeout crew removed the galoshes that kept us from tracking dirt into the spacecraft. Inside the cockpit, everything was pointing up at the sky, so I had to crawl across the ladder, rather than up, in order to get to the flight deck and my seat, which felt like it was hanging off the ceiling. I managed to haul my right leg over the stick, then pull myself up and shimmy into position on the parachute, an uncomfortable bulk under my back. The closeout crew guys, including my friend and astronaut classmate Dave Brown, strapped us into our seats as tightly as they could and helped us get all our connections hooked up—comm, cooling, and oxygen.

We were positioned on our backs for launch, with our knees above our heads, looking straight up at the sky. We were happy to be in our spacecraft, but the position was uncomfortable, especially once we were tightly strapped in.

The preparation for launch was one of the busiest times for the pilot. I was responsible for getting many of the systems ready prior to flight, which meant configuring switches and circuit breakers, starting motors and pumps, and connecting electrical circuits. I configured the reaction control system and the orbiter maneuvering system (the engines that allow the space shuttle to propel itself in orbit). There were many ways I could screw things up so we wouldn't be able to go to space today, and there were many ways I could screw things up so we'd never go anywhere again. Of course, it was possible to throw the right switches but throw them in the wrong order. (People have even screwed up by failing to throw a switch decisively enough.) I learned to follow the checklists precisely, even when I felt I already knew them, because I needed to be so careful—but not so careful that I got behind the timeline, because if certain things weren't in the right configuration by a given point in the countdown, the launch wouldn't proceed. When we were busy, the countdown seemed to go very quickly, but in idle moments it slowed to a crawl.

The countdown clock stopped for the T-minus nine hold. The space shuttle, fully fueled with cryogenic liquid, creaked and groaned. Soon this sixteen-story structure was going to lift off the Earth in a controlled explosion. For a moment I thought to myself, *Boy, this is a really dumb thing to be doing.*

I had been told that astronauts flying in the space shuttle had a risk of death similar to that of Allied infantrymen on D-day. I knew how the crew of *Challenger* had died, and I understood that I was now taking the same risks. I wasn't scared, but I felt aware of the dangers, all at once.

We had been waiting several hours by this point, long enough for some of us to have to use the diaper we wore under our pressure suits. (When the first American to go to space, Alan Shepard, was waiting to launch, a number of technical delays forced him to wait so long that he needed to use the bathroom. He was told to simply go inside his pressure suit, so the first American to leave the Earth did so with wet pants. Ever since, most astronauts have worn diapers or a urine collection device.) Eventually the countdown clock reached the last minute. At thirty seconds, the space shuttle computers took over the launch count. At six seconds, the three main engines roared to life with a million pounds of thrust, but we didn't go anywhere because the shuttle was bolted to the launchpad by eight giant bolts. At zero, the solid rocket boosters ignited and the bolts were exploded in half, setting the shuttle free. We leaped off the launchpad with an instantaneous 7 million pounds of thrust. I knew from watching videos and from seeing launches in person that the shuttle appeared to rise very slowly at first. Inside, though, there wasn't a thing about it that felt slow. One second we were sitting on the launchpad, completely still, and the next we were being hurtled straight up faster than would have seemed possible. I was strapped into a freight train gone off the rails and accelerating out of control, being shaken violently in every direction. We went from a standstill to faster than the speed of sound in less than a minute.

There wasn't much for the commander and pilot to do at this stage other than monitor the systems to make sure everything was going as it should and be prepared to respond if it didn't. People sometimes mistakenly imagined that we were "flying" the shuttle, that our hands were on the controls and that we could move *Discovery* around in the sky if we wanted to, like an airplane. In fact, as long as those solid rocket boosters were burning, we were all essentially just along for the ride. The boosters can't be throttled or shut down.

Once the solid rockets dropped off, two minutes after we left the launchpad, we were flying on the power of the three main engines, so there was more we could do to control our fates. We continued to monitor all the systems closely as we traveled higher and faster. For the first two minutes, we were prepared for the possibility that if something went seriously wrong—most likely a main engine failure—we could turn around and land at the runway at the Kennedy Space Center. We called this abort mode "return to launch site," and it required the shuttle to fly Mach seven backwards. No one had ever tried this and no one wanted to. (John Young, when he was preparing to command the first shuttle launch, said he hoped never to attempt an RTLS because it "requires continuous miracles interspersed with acts of God.") So we were all happy when we got to the point known as "negative return," when RTLS was no longer a possibility and we had other, less risky abort options.

As the shuttle burned through its propellants, it got lighter, increasing its acceleration. When the acceleration got to 3 g's, it became difficult to breathe, the parachute and oxygen bottles I wore on my back in case of emergency pulling on the straps on my chest. The engine throttled back to keep from exceeding the structural integrity of the spacecraft.

As we accelerated, Curt and I, with Billy Bob's assistance, monitored the performance of all the systems on our three cathode ray tube displays, keeping abreast of the procedures so we could be ready at a split-second's notice if we needed to perform one of the actions available to us.

When the shuttle reached its intended orbit, the main engines cut off—MECO—then the now-nearly-empty external tank separated to burn up in the atmosphere. MECO was a great moment because it meant we'd survived the launch phase, one of the riskiest of our entire mission. We had accelerated from zero to 17,500 miles per hour in just eight and a half minutes. Now we were floating in space. I looked out the window.

I tapped Curt on the shoulder and pointed outside. "Hey, what the hell is that?" I asked him. (I was about to use even stronger language, but I didn't know whether we were still being recorded.)

"That's the sunrise," said Curt.

An orbital sunrise, my first. I had no idea how many more of these I was going to see. I've now seen thousands, and their beauty has never waned.

I had been so focused on what we were doing I hadn't bothered looking out the window until now. Even if I had, we had launched in the dark, and up here it was still dark; the sun was behind the Earth. As we crossed over Europe, I saw a blue-and-orange line out the window that spanned the horizon as it grew larger. It looked to me like brilliantly colored paint brushed across a mirror right in front of my eyes, and I knew right then and there that Earth would be the most beautiful thing I would ever see.

I unstrapped myself from my seat and floated headfirst through the passageway to the mid-deck, savoring the alien sensation of weightlessness. When I got there, I found two guys with their heads in puke bags. They were experienced astronauts, but some people have to reacclimate to space every time they go. I'm very lucky that I don't suffer from the debilitating nausea and vertigo that some people do.

On our second full day in space, we reached the Hubble Space Telescope. It's in a much higher orbit than most satellites we might rendezvous with—150 miles higher than the space station. Hubble's orbit is so high, in fact, that missions to rendezvous with it are riskier than flights to a lower orbit.

For many stages of the flight, Curt was in charge of the shuttle controls as commander, and I was there as his backup. But during the rendezvous with Hubble, at a certain point he moved to the back of the shuttle to start monitoring our approach from the aft piloting station and to prepare himself for the manual flying phase. He was to eyeball the closing distance and communicate with me about how we were doing, while I was to make sure we proceeded through the checklist and executed the remaining rendezvous burns properly.

The two spacewalking teams and the robotic arm operator (Billy Bob) moved into high gear once we were safely in orbit. I helped them out when needed and took pictures of Hubble for study on the ground later. Billy Bob was always excited about what we were doing, always enthusiastic, and always had time to help me out or to just take the time to enjoy space. Not everyone who gets the chance to go to space does. He acted as a mentor to me on the mission and taught me all the little details about how to live and work in space that they can't really teach you on the ground, like moving around in zero g, organizing your workspace when everything floats, and of course fun things like peeing while upside down—lessons I would pass on to others as I became more experienced.

Billy Bob was also not above pranking me. I was still the rookie, after all. When I went into my clothing locker to get changed, I discovered that I had only one pair of underwear for the entire mission. Billy Bob had hidden the rest. I think he expected me to panic, but the joke was on him; I didn't really care. He eventually told me about his prank. In retrospect, wearing the same underwear for days was good training for my year in space.

Once we got to orbit, I had to adjust to living in such small quarters

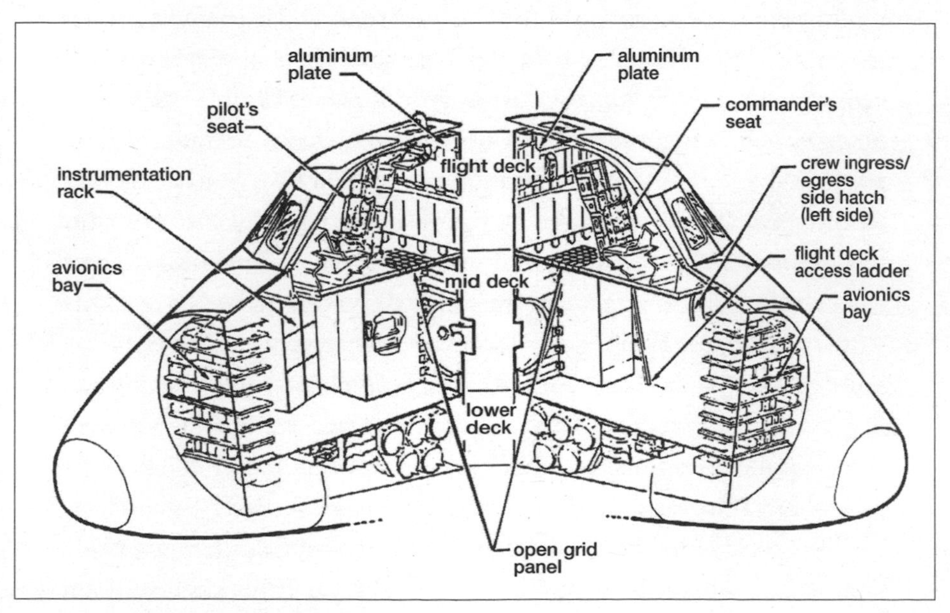

Rendering of the space shuttle cockpit

with six other people. There were two "floors" in the shuttle, the flight deck and the mid-deck, and each of them was smaller than the interior space of a minivan. We worked, ate, and slept on top of one another. At least our eight-day mission would be one of the shorter ones; the longest space shuttle mission was seventeen days.

ONE THING that surprised me about living in space was that it was hard to focus. There were many activities I had done over and over in the simulator, but when I got to space I found it much harder to concentrate on what I was doing. At one level I think it was just the experience of being in space for the first time—who could concentrate on a checklist of a procedure while floating with the beautiful Earth turning just outside the window? At another level, doing basic tasks was much more challenging in weightlessness, and I learned there was no way to compensate for that except to plan for the fact that everything was just going to take a bit more time.

There were physical effects too. Feeling the fluid in my body redistribute itself to my head for the first time was odd and at times uncomfortable. All astronauts experience some level of difficulty concentrating on a short mission—what we call "space brain"—and I was no exception. After you've been in space for weeks or months, you adjust and are able to work through the symptoms, which can vary based on CO_2 levels, vestibular symptoms, sleep quality, and probably other factors too. I couldn't afford to let my work suffer, because there would be serious consequences if I screwed something up.

One of the first things we did when we got to orbit was to open the shuttle's huge payload bay doors. These needed to be open within the first few orbits, in order to keep the electrical systems cool. We needed to deploy and check out the robot arm, or we wouldn't be able to grapple Hubble. If we failed to deploy or activate the Ku-band antenna, we wouldn't be able to communicate with Earth or to rendezvous easily with the telescope. Even tasks like using the toilet required our full attention—I was acutely aware that it was possible to damage it, potentially even permanently, which would mean a premature return.

On day three, Steve Smith and John Grunsfeld conducted their first spacewalk, successfully replacing the gyroscopes. The following day, Mike Foale and Claude Nicollier performed their spacewalk, replacing Hubble's central computer and a fine-guidance sensor. On day six, Steve and John went outside again, this time to install a transmitter and a solid-state recorder. There had been a fourth planned spacewalk, but it was canceled in order to get us back on Earth before Y2K.

Day seven of the mission, the next-to-last day, marked the first time a space shuttle would be spending Christmas in orbit (and, it turns out, the last). We deployed Hubble, and after accepting congratulations from the ground for our success, Curt decided it was time to make his Christmas speech to mission control. He took a piece of paper out of his pocket, cleared his throat, and spoke in his most formal voice into the microphone:

> "The familiar Christmas story reminds us that for millennia, people of many faiths and cultures have looked to the skies and studied the stars and planets in their search for a deeper understanding of life and for greater wisdom . . . We hope and trust that the lessons the universe has to teach us will speak to the yearning that we know is in human hearts everywhere—the yearning for peace on Earth, good will among all the human family. As we stand at the threshold of a new millennium, we send you all our greetings."

Coming from someone else—from Billy Bob, say—this speech might have seemed heartfelt and even moving, but Curt wasn't an emotional guy. As it was, we all sneaked looks at each other. If nothing else, Curt's speech was remarkable for managing to completely avoid any religious content. Maybe Curt was thinking about the time the crew of Apollo 8 took turns reading from the Book of Genesis as they orbited the moon on Christmas Eve 1968. It was a beautiful moment enjoyed by many Christians and non-Christians alike, but an atheist group sued NASA for violating the separation of church and state. Nothing Curt had said would give First Amendment purists anything to get bent out of shape about.

There was a long, awkward pause, both inside the cockpit and on the ground. Usually the capcom would thank the commander for his great speech and reiterate that the spirit of humankind was alive in the space shuttle program or something along those lines. Instead, we just heard static. Moments later, the capcom, Steve Robinson, came on and said simply, "Roger, PLT is go for compactor ops."

The schedule called for the pilot (me) to compact the toilet. In other words, someone needed to tamp down the shit.

Later that night, everyone gathered for dinner on the mid-deck. Billy Bob showed me some special French gourmet food he had brought up with him: quail in red wine sauce, foie gras, tiny liqueur-infused chocolates. No one seemed interested in trying it except me. Billy Bob and I heated it up and took it up to the flight deck. We turned the lights off and played some Mozart, watching the beautiful Earth turning below us while we ate this fantastic food and reflected on how lucky we were to be celebrating Christmas as no one on the space shuttle had done before.

WHEN IT WAS time to go home, I decided to get Billy Bob back for pranking me by hiding the long underwear we layer under the pressure suits for reentry. He didn't suspect anything when he started getting dressed, then he began tossing through his bag of gear again and again, a look of alarm on his face. Once he was thoroughly distressed, thinking he wouldn't be able to get dressed in time for landing, I finally took pity on him.

The landing phase was the most challenging for the commander and pilot. When the space shuttle hit the air molecules of the outer atmosphere at 17,500 miles per hour, the resulting friction created heat of more than 3,000 degrees Fahrenheit. We had to do everything right and trust that the insulating tiles on the space shuttle would protect us.

We did the deorbit burn in the dark at four hundred miles above the Earth. As we moved into sunlight, we seemed troublingly low over Baja California. We had dropped from four hundred miles to just fifty miles entirely in darkness. Curt joked, "We are so low it looks like we won't make it to Florida."

"But we have a lot of smack," I responded. We were still going Mach 25, despite our low altitude.

For about twelve minutes, hot ionized gases built up around the spacecraft. We heard an alarm: one of the air data probes, an instrument that measured air pressure and provided data for controlling the orbiter in the atmosphere, had failed to deploy. This was an emergency, but a minor one, as there were two probes and the other had deployed correctly. Curt and I, with the help of Billy Bob, responded just as we had to this kind of malfunction in the simulator, assessing what had gone wrong and deciding how to proceed safely. In some ways, it was a good thing to have to respond to an alarm like this one. It gave me confidence in the training we had received, that we would be able to handle anything that came our way.

Once we got farther down into the atmosphere and the air became thicker, the space shuttle's airplane design became crucial. Up to that point, it could have been shaped like a capsule, but now Curt was going to land this spaceship in the dark on a runway at the Kennedy Space Center. The space shuttle was a difficult aircraft to land, all the more so because it had no engines that would allow us to pull up and come in for a second attempt. While Curt was at the controls, I had a lot of responsibilities as pilot, a role similar to the copilot of an airplane—monitoring the shuttle's systems, relaying information to Curt, and deploying the drag chute.

I armed, then extended the landing gear at the right moment, and soon after we heard another alarm: a tire pressure sensor was warning us that we might have a blown tire. The space shuttle's tires were specially designed to survive launch, a week or two orbiting in a vacuum, and supporting a heavy vehicle landing at incredibly high speed. If one of them had blown out, our landing could be a disaster. As the alarm kept sounding, I encouraged Curt to ignore the tire pressure—there was nothing we could do about it, and he needed to focus on the landing. I said, "I'll tell you if the next alarm is something different."

He nailed the landing, the tires held under us, and we rolled to a stop. "Nice landing!" I told him, completing one of my most important responsibilities of the whole mission. Our mission was over.

I was surprised by how dizzy I felt being back in Earth's gravity. When I tried to unstrap myself from my seat and get up, I found I nearly couldn't move. I felt like I weighed a thousand pounds. We climbed from the space shuttle to a converted motor home where we could change out of our launch-and-entry suits and get a brief medical examination. Trying to get out of the suit worsened my dizziness, and the world spun up like a carnival ride.

Some of my crewmates were worse off than others, their faces pale and clammy. We were taken back to the crew quarters at Kennedy, where we were able to shower before meeting up with our families and friends. I went out that night to Fishlips, a seafood restaurant in Port Canaveral, with everyone who had come for my landing, and it was a bit surreal, sitting at a long table drinking beer and enjoying fish tacos, when just a few hours earlier I had been hurtling toward the Earth at a blistering speed in a 3,000-degree fireball. We threw a party for our Houston friends when we returned home the next night, and a couple of days later I was back in the office, a real astronaut.

13

September 4, 2015

Dreamed the new people came up here, bringing our total to nine. We were so overcrowded we had to share our CQs. I was sharing mine with some guy I didn't know, and he was cooking meth inside. I had to sleep with a respirator on. The other crew members were getting suspicious of the yellow cloud of smoke coming from under the door, and for some reason I worked to hide it. My roommate kept saying he was going to stop, but he wouldn't. Eventually I tricked him into the airlock, closed the hatch, and spaced him.

IT'S A RARE OCCASION for a Soyuz to dock without another one having left recently. The Soyuz that comes up today is the one that will be my ride back to Earth six months from now, and its crew will bring our total to nine. I'm looking forward to having some new faces up here, but I'm also concerned about how the Seedra will stand up to nine people exhaling rather than six, as well as the strain on the toilets and other crucial equipment. The overall activity level is going to take some getting used to.

Our new crewmates will be Andreas Mogensen (Andy), Aidyn Aimbetov, and Sergey Volkov. Sergey will be here through the end of my mission and will command the Soyuz that he, Misha, and I will go home on in March, but Andy and Aidyn are here for only ten days, flying this short increment that had been meant for Sarah Brightman.

When she withdrew from the flight very late in her preparations to go, her seat was taken by Aidyn, a Kazakh cosmonaut. The Russian space agency has been promising to send a Kazakh to ISS for a long time, as a gesture in exchange for the use of Baikonur as their launch operations center (in addition to $115 million a year). Aidyn is the third Kazakh to go to space but the first to fly under his country's flag rather than the Russian flag.

When the new guys arrive, Sergey Volkov floats through the hatch first. I know him well from being in the same era of space flyers—I was selected in NASA's 1996 class, and he was selected in Roscosmos's 1997 class, so we were peers. At one point, Sergey had been assigned to the STS-121 crew with my brother, and in preparation for that flight they went on a National Outdoor Leadership School trip. They spent a week in a tent in horrible weather in Wyoming, which cemented a lifelong friendship. I got to know Sergey more when we trained together for our Soyuz descent, but because that was so far in the future, we left most of the training for in flight. Sergey was Misha's backup for the yearlong mission, so when we were in Baikonur preparing for launch, Sergey was there with us too. Sergey says to me regularly, "Please say hi to Mark for me."

Then Andy floats through the hatch. He's an ESA astronaut from Denmark whom I've known for years, a friendly guy with blond hair and a perpetual smile. He grew up all over the world and went to high school and college in the United States. His wife jokes that his English is better than his Danish.

When Aidyn comes floating through last, I'm watching with interest. He pauses in the hatch to give a heroic Superman pose to the camera, Gennady and Oleg holding his sides to steady him. He looks a lot like the people I've known in Kazakhstan, more Asian than European. He is younger than me, forty-three, but seems older (maybe it's the zero g). He started his career as a military pilot, rising to the role of flying the Soviet Su-27 Flanker. Then he was selected as part of the first official Kazakh cosmonaut class in 2002. For all these years he's been waiting to fly, sometimes assigned to missions that fell through, sometimes on hiatus when Kazakhstan could not fund his training and flight. I imagine everyone who has flown in space has felt it was a long journey to get here—it's not unusual for American astronauts to wait many years to fly even after completing astronaut training—but Aidyn truly waited a long time.

From the start, Aidyn seems disoriented up here. He gets lost trying to find his way to the Soyuz and ends up in the U.S. lab module; the next day, he can't locate the Japanese module. I find him looking for the 3-D printer in the U.S. segment, and we try to talk about it. But he has no English, and Russian is a second language for both of us, so our discussion is pretty rudimentary.

TODAY WE HOLD the change-of-command ceremony, so I am now officially the commander of the International Space Station. The capcom on the ground congratulates me on taking over for the next six months, and her words hit me—six months is a long time. I try not to dwell on how long I have to go. I've been up here for so long, and I'm only halfway through.

This morning, I showed Andy the view of the Bahamas from the Cupola. Later in the day, he comes to ask me whether the window shutters need to stay closed. At first I'm confused by his question, because I thought the shutters were open. We go to the Cupola, and it's dark outside, a deep, deep black. I explain to him that we happen to be passing over the Pacific during orbital night with no moon and the lights outside the space station turned off for some reason.

In the morning, Gennady greets me, "Good morning, Comrade Commander," with great affection in his voice. I'm going to miss him next week when he's gone—he has been a great commander, and I have learned a lot from him.

Today is Friday, and because there are so many of us we eat Friday dinner in Node 1 rather than trying to cram into the snug Russian service module. Andy has brought us some corned beef and cabbage, which hits the spot; I've been craving a corned beef sandwich from the Carnegie Deli in New York for a long time. After we're done eating, Andy hands each of us a Danish chocolate, an unexpected treat. When we start opening the chocolates, we find that each of them contains a message from someone we know—my chocolate has a poem from Amiko. It was a great idea of Andy's and a really thoughtful gesture.

Football days wet grass cold skinny dips
Foot massages sweet and sour dirty lips
Soft towels home cooking little strings
Burgers and buns no more pipe dreams
Thunder rolls blind folds and fast cars
Palm scratches loamy smells distant stars
Road trips minute beers breezy nights
Real slow dances in pin-striped tights
Sunset warm sand and callipygian
Hot sauce—a lot or just a smidgen
Early morning dew and fireside chats
Enjoy your secret chocolate snack
I'll give you something sweeter
When you come back

On Sunday we have a traditional Kazakh meal, irradiated and packaged into space-food servings: horse meat soup, cheese made of horse milk, and horse milk to drink. The horse meat is a little gamey, but I eat all of it. The cheese is really salty, which is actually a nice change from the low-sodium food we generally have. I comment that the horse milk is really sweet—as commander, I feel like as a gesture of goodwill I should try everything—and Aidyn tells me that it's closest in taste to human breast milk. That does it for me. Now my concern is what to do with a nearly full bag of unpasteurized horse milk. I tell Aidyn I'm going to put it in the small fridge along with the condiments and some science experiments and drink it in the morning with my breakfast. When he isn't looking, I triple-bag it and dispose of it in a spot reserved for the smelliest items.

The next day, I'm floating down to the service module to talk to the cosmonauts when I find Aidyn in the passageway between the Russian and U.S. segments, wedged into a crevice between some hardware stowed on the floor, reading a Russian car magazine. I grab him and say, "Come with me."

I lead him down to the Cupola and show him how to open and close the window shutters.

"You're more than welcome to come down and hang out in here anytime," I tell him. This is a view he would only have a very limited chance to enjoy.

Unlike Aidyn, Andy is very busy. The European Space Agency has sent many science experiments with him. I feel bad for him, because he is spending most of his time on his own in the European Columbus module, which is windowless. I check in with him often to see if he needs any help, and he always seems to be doing well. When Andy isn't working, he can often be found hanging out with us, watching TV or chatting. I encourage him to spend time looking out the window, but I get the sense he wants to be part of the crew just as much as he wants to enjoy the view. I want to say that he's only up here for ten days, so he should be spending all of his free time with his face pressed to the window, but I don't want to tell him what to do. On a ten-day shuttle mission, everyone would be hanging by the windows as much as possible, oohing and aahing.

As much as I enjoy having new faces up here, we definitely feel the strain of having such a full house. With NASA's permission, Sergey sleeps in the U.S. airlock. Without asking permission from the Japanese space agency, I let Andy sleep in their module, since I don't want him to have to spend all his time in the windowless Columbus module. Aidyn sleeps in the habitation module of the Soyuz they will be going home on.

Near the end of Andy's ten-day stay, he remarks, "Boy, do I need a vacation."

"You know what?" I say. "You're complaining to the wrong guy."

He gets it and laughs at himself.

A few days later, I give myself a flu shot, the first one administered in space. We are safe from infectious illness up here, so the shot isn't to protect me; instead it's part of the Twins Study comparing Mark and me. He will be injected with the same serum at the same time—in fact, he insists on injecting himself as well—and then our immunological responses will be compared. When we both tweet about our flu shots, the response is surprising. I even get retweeted by the Centers for Disease Control and the National Institutes of Health. Just the fact that I injected myself seems to be the subject of fascination. I'm learning that sometimes it's the more mundane aspects of life in space that capture the public's attention the most.

On September 12, we gather to see off the short-duration crew. As always, I find it strange to say good-bye to people leaving space. The bond we form up here, a bond of shared hardship, risk, and extraordinary experiences, is powerful. Gennady has prepared the Soyuz, and the crew is suited up in the underwear that go with their Sokol suits. We set up the cameras for the ground to watch as we gather in the service module, then make conversation as we wait awkwardly for the clock to tick down. When it's finally time for them to float through the hatch into the Soyuz, I hug each of them good-bye, especially Gennady. I tell him how much I'm going to miss him. When they are all in the Soyuz, I float in after them and joke that I'm going to stow away. "I'm done, guys. I've decided I'm going back with you!" Everyone laughs as I float back into the station.

We close the hatch, and a couple of hours later they are gone.

Three days later, I hit the halfway point of my mission.

14

As my life was returning to normal following my first spaceflight, in early 2000, I also had a moment to take stock of where I was in my career. What would come next? I had been working for most of my life to become one of the few people who get to travel in space, and now I had done it. I had performed well, our mission had been a success, we had come back safely, and I couldn't wait to go up again. But I didn't know when that would be.

One of my crewmates on the mission I had just finished, Mike Foale, had flown a mission on *Mir,* so he spoke Russian and was well connected within the Russian space agency. He was also an associate administrator of the Johnson Space Center and was close with the center's director, George Abbey, so he had influence with him. Soon after we came back from our mission, NASA was looking for a new director of operations (DOR)—an astronaut who lived in Star City, just outside Moscow, and served as a liaison between the two space agencies. The DOR dealt with the details of training American astronauts to fly on Russian spacecraft and served as the on-site leader for the U.S. astronauts training there. The International Space Station was still in the early stages of construction, and we were ramping up to train international crews in Houston and Star City, as well as in Europe and Japan. Mike said that Mr. Abbey wanted me to serve as DOR. I was flattered, but I was reluctant to take the job. I thought of myself as a shuttle astronaut, a pilot, not a space station guy. I remarked to my brother in private that I didn't want to get that space station stink on me, thinking it would be hard to get off, resulting in fewer shuttle flights.

Still, when I was offered the job, I accepted it. My approach to an unwanted assignment had always been to express my misgivings and my preferences, but then if I was still asked to take the hard job, I did my best to make it a success. I was to start just a few months later.

Mike flew with me to Russia the first time to help me get acclimated. We were met at the airport by a Russian driver named Ephim, a squat, gruff bull of a man. I would later learn that Ephim would do anything to protect us and our families, even physically if required, and he cooked a great *shashlik,* Russian barbecue. Ephim loaded us into a Chevy Astro van, one of the few Western vehicles in Russia at the time, and I watched Moscow go by as we passed through the city. The snow was piled high, and the car exhaust and other pollutants had stained it dark. As we traveled northeast from Moscow, past old Russian cottage-style houses with their ornate trim and elaborately shingled roofs, the snow gradually turned white. Soon we were passing through the gates of Star City.

Down a narrow path lined with thick birch trees, past old Soviet-style cinder-block apartment buildings and the giant statue of Gagarin holding flowers behind his back and leaning forward welcomingly, we arrived at the awkward row of Western-style town houses built for NASA we called "the cottages." It was Friday night, so after dropping our bags we went straight to Shep's Bar, actually just the remodeled basement of Cottage 3. The place was named for Bill Shepherd, a NASA veteran of three space shuttle flights who was now in Star City training to become the first commander of the International Space Station. He was also a former Navy SEAL who was legendary for saying in his astronaut interview, when asked what he could do better than anyone else in the room, "Kill people with my knife." Bill had a penchant for putting people under the table in a drinking game called liar's dice, and my first night in Russia I was expected to participate. I wasn't one to argue, and I even had a slight advantage over the others in that I had played the game in my fighter pilot days. Shep had no mercy on us newcomers, though, and I watched as some scientist astronauts who were in Russia for the first time fell out one by one. Shep didn't need a knife to kill; he could also kill with dice.

Even though I held my own, the next morning was rough when I had to get up very early for a four-hour ride on a bumpy road in a bus smelling of burning engine oil. I lay down on the backseat and tried to sleep as we headed to Russa, the remote village where space flyers trained in case the Soyuz landed in cold weather. The plan was for me to first observe, then to participate in, the Russian winter survival training.

During the reign of Ivan the Terrible, Russa had been a thriving city, but now, having been largely destroyed in World War II, there wasn't much there aside from a "sanatorium," a quintessentially Russian combination of hospital and hotel that to Americans looked more like an old spa. The area is famous for spring-fed lakes that were supposed to have healing powers.

Unbeknownst to me, and against NASA's objections, I was to go through the same psychological evaluations the cosmonauts did, and this was the first order of business on my first day. NASA had its own psychological evaluation process, of course, but the Russians' was a bit different. The first test I did involved sitting across from a psychologist under a bare bulb, both of us sitting on hard wooden kitchen chairs. I felt as though I were going to be interrogated like Francis Gary Powers during the Cold War.

The psychologist, who looked like a well-fed version of Sigmund Freud, explained the test: I was to estimate various lengths of time by stopping a stopwatch without looking at it after what I thought was ten seconds, then thirty seconds, then one minute. I took the stopwatch from him and held it down by my side to begin the first test. I soon realized that I could see the doctor's watch from where I was sitting, including the second hand. I "estimated" each of the intervals of time perfectly. The psychologist reacted with shock and congratulated me profusely on my time-estimation prowess.

Once the test was over, his watch was no longer visible to me, and I wondered whether that had actually been a test of my honesty, or perhaps a test of my ability to adapt. I decided not to worry about it much—to me, using any available tool I had to excel on the test was at least as important as following the rules blindly. I don't condone cheating, but I've learned it's important to be creative in solving problems. Now that I've gotten to know the Russian culture, I think my approach was the right one.

After spending a few days sharing a dank room with a NASA flight surgeon who was monitoring the training of the previous crew, I joined American astronaut Doug Wheelock and cosmonaut Dmitri Kondratyev as a three-man crew. I didn't know yet that I would wind up flying in space with both of them much later in my career. Doug was an Army officer and helicopter pilot, even-tempered and easy to get along with. Dima was a fighter pilot who had flown the MiG-29, one of the people I might have wound up in air-to-air combat with at an earlier stage in our lives. In fact, years later we figured out that we were once stationed on opposite lines of the Soviet border in Scandinavia, him protecting the Russian Bear bombers and me in the F-14 Tomcat protecting the carrier battle group.

The survival training was grueling. We were sent out to a field with a used Soyuz capsule to simulate a remote landing, equipped with nothing but the emergency supplies carried in the spacecraft. Dima didn't speak much English, and neither Doug nor I spoke great Russian, but the three of us communicated well enough to get through the training. We built shelters, made a fire, and tried to keep from freezing to death while we awaited "rescue." It was so cold the first night we were unable to sleep, so we stood in front of the fire, slowly rotating in order to keep any side from getting too cold. In an uncharacteristic act for a Russian, Dima broke with protocol and at five a.m. announced we would build a teepee in order to stay warm. Cutting down trees with a machete in the freezing dark winter night was miserable, but by seven a.m. we had our shelter assembled out of birch limbs and the Soyuz parachute. We were now able to keep warm, though the teepee quickly filled with smoke. We kept our heads as low as possible so we could breathe as we slept.

On the last day, we hiked through the woods, a navigation exercise to simulate meeting up with rescue forces. The landscape was stunning, with stands of birch trees stark against the sky, everything covered with a fresh layer of fluffy snow, the new flakes sparkling in the morning light. We emerged from the forest onto a large frozen lake that was steaming in the subzero temperatures, dotted with old Russian men sitting on their pails, ice fishing. This image struck me as serene and quintessentially Russian. Seemingly frozen in time, like an epic scene from the film *Dr. Zhivago*, it was a moving sight that will be etched in my memory forever.

IN MAY I moved to Russia to start my position as DOR. It was a big transition. NASA and Roscosmos were in the process of figuring out how to train international crews together to work on an international space station, a huge undertaking with a lot of potential for power struggles, cultural conflicts, and temper tantrums from big egos on both sides. But I liked the job in Star City and found it easy to settle in. I lived on the eighth floor of one of the cinder-block Soviet apartment buildings, and each day I walked the path from my apartment, past Gagarin's statue, past the town houses the U.S. astronauts lived in while training for flight, to the profilactorium (or "profi," as we called it), the Star City cosmonaut quarantine facility where NASA had also been given offices.

I found it challenging at times navigating the issues between the Russians and Americans. We had different languages, different technology, and different ideas about the best way to fly in space. But I liked the Russians I met and took a real interest in their culture and history, building the foundation for our future collaboration on the ISS.

The first module of the International Space Station, the FGB, had been launched from Baikonur in November 1998, followed two weeks later by Node 1, the first U.S. module, which launched on space shuttle *Endeavour*. When the two were joined together, it was a major international accomplishment. The infant space station wasn't ready to be permanently occupied, though, because it lacked necessary features like a life support system, a kitchen, and a toilet. It orbited empty for the next year and a half until the addition of the Russian service module, which made it habitable.

Leslie and Samantha came to join me in Russia for the summer.

In late October 2000, I traveled to Baikonur for the launch of Expedition 1, the first long-duration mission to the ISS. Bill Shepherd would be launching on a Soyuz with two Russian cosmonauts, Yuri Gidzenko and Sergei Krikalev. This would be only the second time an American was traveling on a Soyuz. Another three-person crew would be replacing them in March, and it was hard to believe that the station would be occupied nonstop from then on. Since I still thought of myself as a space shuttle guy, I didn't assume I would fly a long-duration flight on station myself—I hoped to be assigned to another shuttle mission soon, as pilot again. Then if I was lucky I might fly two more space shuttle missions as commander, and that would probably be the end of my spaceflight career. Having spent a total of eight days in space, I found it impossible to imagine that I would live on the space station one day, let alone set records there.

The night before the Soyuz launch, there were celebrations and the traditional toasts and revelry. A NASA manager in town for the event had more than his limit—way more—and I spent the day taking care of him because he was too ill to be left alone. The next morning I saw Shep briefly when he was on his way to get suited up for launch.

"What the fuck was going on last night?" he asked me. "It was like a fucking frat house with people yelling and screaming and banging on my door. I barely got any sleep."

"Sorry about that, man," I said. "Good luck in space."

The Soyuz launched safely that day, but I didn't get to see it—I was busy helping my naked colleague vomit in the bathtub. I was sorry to miss the launch, but happy to be in Baikonur on this historic day. I was enjoying living and working in Russia more than I had expected. I watched on television in the old cosmonaut hotel as the spacecraft disappeared into a tiny point in the sky; I had no idea how much of a role the Soyuz, and this place, would play in my future.

Soon after I came back from Russia the following year, Charlie Precourt, the head of the Astronaut Office, asked me to serve as backup to Peggy Whitson for Expedition 5 to ISS (the fifth expedition of overlapping crew members), to launch in June 2002. Normally, the backup crew would fly two expeditions later, so their service flows naturally from their backup training to their flight. Because of unusual circumstances, I wouldn't be on the upcoming flight, so serving as backup would be a pretty shitty deal. My first reaction was to decline. A mission to the International Space Station was very different from what I had trained for and, to a certain extent, from what made me want to be an astronaut in the first place: test piloting a rocket ship.

"If I'm being honest, I'm not sure whether I ever want to spend six months on the space station. I'm a pilot," I told Charlie. "I'm not a mission specialist. Science really isn't my thing."

Charlie understood; he was a pilot too. He explained that he hadn't been able to get anyone to agree to serve as Peggy's backup, having gone through most of the more experienced astronauts. He offered me a deal: if I would serve as Peggy's backup, which would mean returning to Russia for a significant period of time to train on the Russian ISS systems and on the Soyuz, he would assign me as commander of the space shuttle on my next flight, and as the commander of the International Space Station after that. After giving it a lot of thought, I went into his office with a list of reasons why I still thought I was the wrong person for the job. Charlie listened patiently.

"All that said," I told him, "I've never said no when someone asked me to do something hard. So if you ask me to do this, I won't say no."

"I'm not going to accept that," Charlie answered. "You're going to have to say yes."

"Okay," I said somewhat grudgingly. "Yes, I'll do it."

I had been given this assignment later than normal, so in addition to taking a job that didn't feel natural to me, I was trying to play catch-up. I trained a great deal in Russia, learning their Soyuz and the Russian part of the ISS. I also worked to hone my skills in the Russian language, which I had always found excruciatingly difficult. In addition to this, I had to learn the U.S. segment of the space station, which is incredibly complex; how to fly the space station's robotic arm; and how to do spacewalks.

I went through Russian water survival training with Dima Kondratyev, whom I had gone through winter survival training with, and cosmonaut Sasha Kaleri, my two new backup crewmates. We left early in the morning on September 11, 2001, on an old Russian Navy vessel from Sochi, a palm-tree-covered coastal town on the Black Sea at the base of the Caucasus Mountains. As we slowly motored out to sea, we were given a tour of the ship and shown how to use some of the equipment. Toilet paper was forbidden, as it clogged up the sanitation system. We were told instead to use a brush soaking in antiseptic next to the toilet. *Community ass brush?* I thought to myself. *Shit!*

The water survival training wasn't much more pleasant than winter survival training—an old Soyuz was lowered into the water, and we had to climb into it wearing our Sokol launch and entry suits. The hatch was closed behind us, and we sat there in the stifling heat until we were directed to remove our Sokol suits and put on our winter survival gear, followed by a rubber anti-exposure suit. It was almost impossible to follow these directions in the tight confines of the Soyuz. Dima, Sasha, and I had to take turns one by one lying spread out across one another's laps to struggle out of one suit and into the other. The capsule heaved up and down with the rolling swells of the Black Sea, and I thought about how impossible this would be if we were returning from space and already weakened from living in zero gravity. Once in my winter clothing—not pleasurable since the Soyuz was as hot as a sauna—I then had to put on the full rubber anti-exposure suit, including layers of hats and hoods. We were drenched in our own sweat and exhausted even before climbing out of the Soyuz and jumping into the sea. This wasn't really about training on the hardware or learning techniques; like winter survival training, it was almost exclusively a psychological and team-building exercise in dealing with shared hardship. To me, it would have been more effective to just admit that fact.

Once we finished up our training, we headed back to the bridge of the ship, where the captain toasted our success with vodka. I reflected on how strange this scene would have looked even just a few years before—me, an officer in the United States Navy, drinking alcohol on the bridge of a Russian Navy ship with its captain and Dima, a Russian Air Force pilot.

As we got back on shore, we got a call from Star City telling us that two planes had just crashed into the towers of the World Trade Center. We were as shocked as the rest of the world, and for me it was a horrible feeling to be so far from my country when it was under attack. We found the nearest television, and like most people at home, I spent hours watching the coverage and trying to understand what was happening. The Russians rose to the occasion, doing everything they could to help us. They brought food, translated the Russian news so we could understand what was going on, and even canceled the remaining training to get us back home as soon as possible. We flew out of Sochi the next day, and I was startled by how much the security had increased at the airport, despite the fact that the terrorist attack had been in another country on the other side of the world. As we waited in Moscow for flights to the United States to resume, we saw flowers piled high outside the gate of the U.S. embassy in a show of solidarity that I will never forget.

While in Russia, I also got to spend time with the prime crew—Peggy Whitson, my classmate, as well as Sergei Treshchev and Valery Korzun. Valery, who would be the commander of Expedition 5, was an atypical Russian with a welcoming smile and an endearing personality.

As part of our training, we had to learn to fly the Canadian robot arm, so Valery and I traveled together to Montreal in one of NASA's T-38 jets. This was a rare opportunity to fly in a T-38 for a Russian cosmonaut, and it was fun for me as well to fly with a former Russian fighter pilot. After we completed our training in Montreal, I wanted to stop at my old Navy base, Pax River, for the annual test pilot school reunion. There I could catch up with old friends like Paul Conigliaro, and I thought Valery would enjoy meeting some Navy test pilots and they him. I made sure to get the appropriate permission before landing on a U.S. Navy base with an active-duty Russian Air Force colonel. I also had to make sure a U.S. customs official would meet our plane, since we would be flying directly from Canada.

When we landed and parked on the tarmac, right next to the Chesapeake Bay, the customs official wasn't there yet. When I called, he said he hadn't left his office—ninety minutes away in Baltimore. He told me sternly that we were not to leave the airplane until he arrived, but it was below freezing and windy, and Valery and I were wearing only our NASA blue flight suits and light flight jackets. I told the customs official we weren't going to freeze to death waiting for him and would be in the Officers' Club and hung up while he was still yelling at me to stay at the airplane. Had we had the proper supplies, perhaps we could have constructed a teepee.

We proceeded to the bar and spent the next couple of hours by the keg, sharing airplane stories. Valery told us about what it was like being a Russian fighter pilot and cosmonaut and charmed my former Navy colleagues. Eventually, the customs officer barreled into the O Club, telling everyone who would listen that he wanted to take Valery and me to jail for violating his orders. The commanding officer knew me from my previous tour as a test pilot and had enjoyed Valery's company, so he told the customs official to do his paperwork and then get off his base. Valery went on to become the deputy director of the Gagarin Cosmonaut Training Center at Star City, and he's had my back ever since.

Peggy's launch went off without a hitch in June 2002, and soon after I was assigned to be the commander of my second space shuttle mission, STS-118, tasked with delivering new hardware to the International Space Station. The mission would be twelve days, and we were scheduled to fly on the space shuttle *Columbia* in October 2003. True to his word, Charlie Precourt had made sure I was assigned as commander, even though he was no longer the chief astronaut.

Since this was only my second shuttle flight, and I hadn't yet been to the ISS, the new chief astronaut wanted my pilot to be someone who had spaceflight experience. That sounded simple enough, but all the pilots who had already flown at least once were my classmates, and generally classmates aren't asked to command one another, especially when they have the same amount of experience. Kent Rominger, the new chief, and I discussed the options. The only pilots not currently assigned to a mission were Charlie Hobaugh, Mark Polansky, and my brother. Of these, I thought my brother was the best fit: we got along (at least since we stopped beating the crap out of each other at age fifteen), we understood each other, and we knew that being classmates wouldn't cause any issues between us. NASA was all for it.

As we got closer to making the assignment official, I thought better of it. The story of identical twin brothers serving as commander and pilot of the same mission would bring an enormous amount of attention. In some ways this would be a good thing, of course—NASA was always looking for ways to engage the public's imagination and get people interested in spaceflight. But I didn't want this flight to be seen as a publicity stunt, and I didn't want the story of twins in space to distract attention from our mission or my other crew members.

Another concern was more personal. Both Mark and I were always aware of the risks we took each time we went to space. For me, the possibility that my daughter might be left fatherless was always offset slightly by the fact that, even if the worst happened, she would still have her uncle Mark in her life as a stand-in father—one who would remind her of me. Each time Mark went to space, I was aware that I might have to play the same role for my nieces. If Mark and I were to fly in space together, we would have to accept the possibility that our children could lose both their dad and their uncle all at once. The more I thought about it, the less I thought it was a good idea.

That left just two candidates: Charlie Hobaugh and Mark Polansky. Polansky wasn't interested in flying as my pilot, since he technically had more experience than I did, having flown to ISS before, which was understandable. That left "Scorch"—Charlie Hobaugh. Scorch had a reputation for being very direct—if he thinks you're wrong, he won't hesitate to let you know. He told me he didn't mind flying with a classmate as his commander. He said he appreciated any opportunity to fly in space, and I knew he meant it.

So my crew was set: Scorch would be my pilot, and the rest of the crew would be rounded out by five mission specialists: Tracy Caldwell, Barbara Morgan, Lisa Nowak, Scott Parazynski, and Dave Williams.

I was most concerned about Lisa, whom I had known longer than most of my colleagues, about fifteen years, since we were in test pilot school together at Pax River. She was a technically brilliant flight engineer. But lately she had become obsessive about small details that didn't seem to matter much, like what she was going to have for lunch that day. She could become hyperfocused and had trouble letting things go, even if they were irrelevant. On Earth this wasn't a problem, but on a spaceflight, every member of the crew was crucial to its success, and these peculiarities of Lisa's personality began to concern me.

On the morning of February 1, 2003, I was standing on my front lawn looking north. It was a Saturday, just before nine a.m., and a shuttle mission with seven of my colleagues, including three of my classmates, was returning to Earth. I thought I might be able to see the streak of fire as *Columbia* entered the atmosphere north of Houston on its way to land at the Kennedy Space Center. It was foggy, but as I watched the sky I saw a bright flash in a break in the fog. *Columbia!* I went back inside and ate a bowl of cereal. As it got closer to the planned landing time, I started paying more attention to the TV. The orbiter hadn't landed yet, so NASA TV was switching between live shots inside mission control and the runway at Kennedy Space Center. I noticed Charlie Hobaugh in the control center—he was acting as capcom that day—and I saw he was slouching low in his chair. That was a strange sight, especially for him; he was generally a squared-away Marine, so slouching on the job was uncharacteristic. I emailed him, half joking, saying that he should sit up straight because he was on TV. Then I heard Charlie say, "*Columbia*, Houston, comm check." A long pause went by. There wasn't an answer. This wasn't normal.

Charlie spoke again. "*Columbia*, Houston. Comm check. *Columbia*, Houston. UHF comm check." He had switched to the backup comm system. Still no response from *Columbia*. My heart started beating faster. The countdown clock got down to zero and started to count up. *Columbia* was supposed to be on the ground by now, and, being a glider, it had little margin to arrive late. Charlie kept making the same call over and over again. I jumped in my car and headed to the space center, dialing my brother on my cell phone. My call woke him up. By then reports were coming in that pieces of the orbiter were falling about a hundred miles north of Houston. Mark and I talked about the parachutes, the possibility that the crew might have survived using escape procedures that were developed after the *Challenger* disaster. Every subsequent shuttle crew trained to extend an escape pole out the hatch, use it to slide out past the wing, then parachute down to safety. No one had actually tried this, of course. Mark and I hoped that it could work, though we weren't optimistic.

It soon became clear what had gone wrong. The space shuttle's external tank, which was sort of like an enormous orange thermos, was covered with foam to help insulate the cryogenic propellant inside and keep ice from forming on the surface. Almost from the start of the shuttle program, the vibration of launch and subsequent air pressure as the vehicle accelerated had been causing pieces of foam to fall off the tank. Engineers had been unable to completely resolve the issue. Usually the foam fell away from the orbiter, or fell in small enough bits that there was little damage. But the day *Columbia* launched, a noticeably large piece of foam, about the size of a briefcase, had fallen and struck the leading edge of the orbiter's left wing, a particularly bad place for the heat shield to be damaged. There had been a brief discussion on the ground as to whether this foam strike would cause a problem, and the managers and engineers involved had quickly concluded that it would be fine. The crew of *Columbia* was never a part of these discussions, and though they were informed of the foam strike, they were told the impact had been analyzed and that there was "absolutely no concern for entry."

Seventeen years earlier, the *Challenger* commission had blamed that disaster on a creeping complacency about safety in the shuttle program. The culture at NASA had changed a great deal as a result, but now it seemed maybe that complacency had crept back in again. It's not as though no one had raised the alarm about this issue: Apollo veteran John Young, commander of the first space shuttle mission, and conscience of the Astronaut Office, was always standing up in our Monday morning meetings, trying to convince people of the danger posed by the foam. I remember him saying distinctly, "We have to do something about this or a crew is going to die."

I thought about the people I knew who had been on *Columbia*. I had known Dave Brown longer than most of my classmates because he had been at Pax River when I was. He had a great gap-toothed smile and a casual attitude that belied his enormous accomplishments—he had been admitted to an elite program that allowed flight surgeons to become Navy pilots. He had helped Mark prepare for his NASA interview and then helped me when I was called. That was the type of guy he was.

Laurel Clark was a Navy doctor before she became an astronaut, and our families had become close soon after we moved to Houston. She had a son, Iain, the same age as Samantha. Laurel would often pick up Samantha and take her along with Iain to the zoo on Saturdays. Laurel and her husband, Jon, were part of an inner circle that met often for social evenings at Mark's house. Laurel liked wine, and so did the rest of our group, and we spent many great evenings together. We gave her the nickname "Floral" for her flowery fashion sense and her love of gardening. She had a carpet of violets at her house, and in the weeks and months after the accident everyone in our class would be given a small pot full of them to care for and remember her by. Most of us kept them on the windowsills in our offices, and Lisa Nowak would often come by and take care of our violets for us if they weren't doing well.

Willie McCool, a fellow Navy pilot, and I had crossed paths briefly at Pax River before we were both selected as astronauts. He had been finishing up his tour as a test pilot when I was just starting mine. I remember the first time I saw his name on a list of the new class and thought it had to be the best astronaut name ever. Willie was infectiously positive, extremely smart, and genuinely caring about the people around him.

I didn't know the other crew members nearly as well because they hadn't been in my class. Rick Husband, the commander, a dedicated family man and Air Force pilot; Kalpana Chawla, the first Indian American woman in space and an aerospace engineer; Mike Anderson, an Air Force pilot with a ready smile; and Ilan Ramon, an Israeli fighter pilot who had been chosen to represent his country on this shuttle mission. Ilan was considered a national hero, the youngest pilot to have taken part in a risky air strike against an Iraqi nuclear reactor in 1981.

He subsequently became one of Israel's first F-16 pilots. The crew left behind a total of twelve children.

In my experience, when colleagues have died in accidents, we find ourselves reflecting on what great people the deceased were. Still, it was a special blow to lose a group of seven people who were all so warm, generous, and kind. It was as though we had lost the seven most respected and well liked of all our colleagues.

That day, my brother and I decided on our own to get some astronauts up to the area where the debris was falling. This was a bit ballsy of us, as we weren't very senior in the Astronaut Office. We called George Abbey, now the former director of the Johnson Space Center, who continued to hold a great deal of sway in Houston. He recommended we call the Harris County constable, who got us in touch with the Coast Guard at Ellington Field. Mark and one of our astronaut colleagues got into a helicopter and were soon searching through the East Texas terrain for debris and the bodies of our friends and colleagues.

I stayed back with a large group working on a recovery plan for the astronauts' remains and the orbiter debris, so we could reconstruct what had happened. After the *Challenger* disaster, pieces of debris recovered from the ocean floor provided physical proof of what had gone wrong, and, as with *Challenger*, we would gather pieces of the shuttle in a hangar at the Kennedy Space Center in Florida. When I got home that evening, Leslie and I went to my brother's house to be with Jon Clark, Laurel's husband, and Iain, now eight. They had just returned from Florida after the horribly long, futile wait at the landing facility. It was heartbreaking to see them and try to comfort them. Our classmate Julie Payette was temporarily staying with Mark and his family at the time, and she and I tried to impress upon Jon and Iain that the crew's deaths were likely painless. We had no way of knowing this for sure, of course, but we wanted to believe it for ourselves as much as for Laurel's grieving family. Later we would learn the crew probably had less than ten seconds of useful consciousness after the orbiter's pressure hull was breached. None of them had time to lower their helmet visors, so we knew depressurization must have taken place very quickly. After one of the control panels was recovered from a field, investigators deduced that Willie had tried to restart two of the auxiliary power units, so we knew they must have had at least a sense that something was going wrong.

The next day, I headed north in my car and helped out with the search for debris and human remains. I was teamed up with an FBI evidence-response team that had been involved in identifying remains at the World Trade Center. They worked with dogs that could distinguish human from animal remains. Standing in a wooded area where debris had fallen, I thought about other airplane crashes that had killed my friends and colleagues. The charred smell, the search for pieces of smashed aircraft, and the burned remainder of an elegant flying machine—all of it reminded me of the opening pages of *The Right Stuff.* In all my years of flying and scores of colleagues lost, this was my first time as part of the accident recovery team, like the pilots in Tom Wolfe's book. I don't think Tom ever saw such wreckage himself, but I could now confirm that he described it all perfectly.

Word had spread at JSC about the search, and a large number of NASA workers volunteered to help. But the area where debris had fallen covered so many thousands of square miles, from central Texas to Louisiana, that we needed more people. Emergency workers from all over the country, many of them Native American firefighters from the western states, descended on the area and quickly set up tent cities, complete with their own supplies. I was impressed by their dedication, organization, and skill at walking detailed search patterns in the thick woods of East Texas. They recovered thousands of fragments of *Columbia,* and every piece would help us figure out what had gone wrong.

At the Kennedy Space Center, workers started to assemble parts on an outline of the shuttle's silhouette painted on the concrete floor of a hangar. The first time I walked into that space to see the debris laid out, I was struck by the sight. The fact that a spacecraft can hit the atmosphere and burn up, yet the pieces can still be identified and reassembled this way, was eerie. I had been assigned to the next flight of *Columbia,* and it was strange to see the orbiter that was supposed to have been mine to command mangled and burned on the concrete floor. I later learned that it had been a toss-up between Willie McCool and me as to who would serve as pilot of my Hubble Space Telescope repair mission and who would fly the ill-fated mission of *Columbia.*

Since the debris field was so large, the pieces of the orbiter couldn't all be recovered on foot. A couple of weeks later, I was put in charge of directing an air search, using airplanes and helicopters to locate the larger pieces. You would think a piece of a spacecraft would be instantly recognizable, even from the air, but we wasted time investigating old cars, bathtubs, rusted-out appliances, and all kinds of garbage that, from a distance, looked like it could have come from the shuttle. There were rumors of remains of murder victims found during the search, and sites the searchers thought looked like methamphetamine labs, though I could never determine whether these rumors were true.

Of the debris we did find from *Columbia,* some of it was strangely undamaged. I found the space shuttle's Canon printer lying in the woods without a scratch on it—the same model of printer that I would later struggle with while living on the space station. We found samples from science experiments the crew had worked on, still intact—so much so that scientists could complete some of the research goals of the mission. A petri dish full of worms even survived the disaster.

Every day I was out searching, the Salvation Army was out there too, providing food and coffee, offering any kind of help they could. Ever since, I never walk by their ringing bells at Christmastime without putting something in the red kettle.

A few of the astronaut doctors worked in the local morgue, safeguarding the remains of our fallen colleagues as they awaited transport. Eventually I escorted Laurel's body from the morgue to Barksdale Air Force Base in a Black Hawk helicopter. As I climbed out of the helicopter, I was surprised to see an Air Force general in full dress uniform saluting sharply, behind him a full formation of officers and airmen at attention. I was moved by their show of respect while the flag-draped casket was carried into the hangar. Later, Laurel's remains were transferred to an aircraft to be flown to Dover Air Force Base in Delaware, the military's mortuary, for a forensic autopsy.

As the search went on, a second tragedy occurred: a Forest Service helicopter crashed while searching for debris. Two people were killed and three more were injured. The ensuing investigation revealed that the pilot was flying outside the operating limits for the aircraft, maybe in an effort to get to a hard-to-reach area. No one talked about calling off the search for debris and remains, but this was another sobering reminder of the risks inherent in aviation.

Three of the crew were buried at Arlington National Cemetery, and other funerals were held in the crew members' home states. NASA hired or borrowed airplanes to take those of us closest to the crew to Arlington and to the other funerals. On a blustery day that would have been Laurel's forty-second birthday, she was laid to rest next to two of her *Columbia* crewmates. Seeing the pageantry of the full military honors, and the finality of her casket lowered into the ground, I absorbed fully the loss we had suffered and became more aware than ever of the risks we were taking traveling into space. I had lost friends and colleagues to airplane crashes many times before. I'd stopped keeping an exact count when the number got into the thirties; it is now in the forties—but I had never lost anyone as close to me as Laurel Blair Clark.

I can honestly say the *Columbia* accident never for a second made me think about quitting. But my colleagues' deaths gave me a renewed sense that my daughter could have grown up without a parent, just as the *Columbia* crew's kids have done. The shuttle program had been suspended until the accident investigation board could come to a conclusion about what had happened, so I didn't have much to do for the next six months. Eventually I was named chief of the Space Station Integration Branch, heading up a group of astronauts and engineers making decisions about hardware and procedures for the International Space Station, which had now been inhabited nonstop for more than two years. (It was still small and rudimentary compared to the expanded station I would visit in the future.) I was learning everything I could about how to make the station work most efficiently and effectively.

In August 2003, the *Columbia* Accident Investigation Board submitted its findings. It did not call for the shuttle program to shut down completely, as some had feared. But it would not be allowed to continue forever. The board recommended that after the assembly of the

International Space Station was complete, planned for 2010, the shuttle orbiters should be recertified in order to keep flying. This process would require dismantling and rebuilding all three orbiters from the ground up. Recertification would be so complex and expensive there was no way NASA would be able to get Congress to pay for it, so we knew that most likely the shuttle would be scrapped. Besides, NASA wanted to focus on a new exploration vehicle (the project that has now become the Space Launch System and Orion) and wouldn't be able to fund it properly while supporting both the space shuttle and the space station. The shuttle program would be the one to go. I agreed with that decision, though I knew I would miss it.

IN OCTOBER 2003, Leslie gave birth to our second child, Charlotte. The birth shaped up to be even more difficult than Samantha's. When Charlotte was delivered by C-section, she had no heartbeat and wasn't breathing. I still remember the sight of her tiny limp blue arm hanging out of the incision, while Leslie's doctor was yelling for help. I'd had a great deal of training and experience dealing with emergencies, but the situation in the operating room was so disturbing I had to leave. My brother and Samantha were in the waiting room, and they told me that I looked as white as a sheet as I came out of the OR. I sat with them for what seemed like an eternity, until Leslie's doctor came out to tell us that both Leslie and Charlotte were fine now, though it had been touch and go for a while. He warned me that because Charlotte had been deprived of oxygen for some period of time during birth, she might have health problems as she grew up, including the possibility she could have cerebral palsy. He had no way of knowing what her outcome might be, and it was his professional responsibility to warn me of the possibilities. But when I asked his personal opinion, he said, "I don't think she'll have cerebral palsy. I think she'll be just fine." He was right.

Our mission was put back on the schedule for September 2006. Not long after, it was postponed to June 2007. All this reshuffling gave me the opportunity to make changes to my crew. I suggested that Lisa Nowak should get to fly on an earlier flight for two reasons: her obsessiveness gave me pause, and if she had to wait to fly on STS-118, it would be nearly ten years after she had been accepted as an astronaut. I argued that she should be put on the second return-to-flight mission, which would fly well before ours. As luck would have it, that mission had my brother, Mark, on it.

At the same time Lisa was moved, Scott Parazynski was moved as well, to the mission just after mine, with Pam Melroy as commander. In exchange for Scott, we got Rick Mastracchio. Rick had worked as a flight controller in mission control before applying to become an astronaut, and in that role he had designed many of the contingency abort procedures we practiced in the simulator. I knew this would make him an invaluable crew member during ascent and entry, and he was extremely competent with everything technical.

Part of being an astronaut involves having your health monitored more closely than most people's. Every year I had my annual flight physical in February, the month of my birthday, and February 2007 was no exception. After my physical, I was told that I had a slightly elevated level of prostate-specific antigen. All men have a certain amount of this enzyme in their blood, and levels can vary naturally, but an elevated level can be an indicator of prostate cancer. Because my levels weren't very high, and because I would be unusually young to be diagnosed with this kind of cancer, I decided to wait until after my upcoming mission was over to investigate it further.

STS-118 was a mission to deliver a number of key components to the International Space Station: a small truss segment, an external stowage platform, and a new control moment gyroscope, a device that allows the station to control its attitude. We were also to carry a SPACEHAB logistics module, which was packed with supplies to bring up to the station. When it returned, it would carry science samples, broken hardware, and garbage back down. We would be flying the sixth mission after the loss of *Columbia,* and several of the subsequent missions had withstood damage to the heat tiles from debris falling during launch. Each time, engineers examined the damage and determined anew how to avoid it, but then it would happen again. I would have preferred that tiles not be damaged, of course, but I was glad the issue was being taken seriously now, and it seemed to me we were doing everything we could to mitigate the risk.

The crew assignments for this flight were now finalized: Scorch, Rick Mastracchio, Barbara Morgan, Dave Williams, Tracy Caldwell, and, late in our training, Alvin Drew.

Barbara Morgan had been an elementary school teacher in Idaho when she was named a finalist for the Teacher in Space program in 1985. When Christa McAuliffe was chosen to teach lessons from space on *Challenger*, Barbara was designated her backup. She trained along with Christa and the *Challenger* crew for the entire year, preparing to complete the mission if for some reason Christa wasn't able to. After the traumatic experience of seeing *Challenger* explode in the sky over Florida with seven good friends aboard, a lot of people would have distanced themselves from that tragedy. But to her credit, Barbara volunteered to go on the national tour that had been planned for Christa after the mission, visiting schools all over the country to talk about the space shuttle and the importance of education. Barbara wanted the schoolchildren to hear from someone who had shared Christa's dream of flying in space and still had faith in the space program. Barbara officially joined the astronaut corps in 1998 and worked in a number of positions before being assigned to her first flight—this flight with me. When she flew in space, it would be twenty-one years after the *Challenger* disaster.

Barbara was also the only astronaut to have been chosen for the corps completely outside the process of the astronaut selection board. For this reason, some of our colleagues regarded her with skepticism. I decided to reserve judgment until I got to know her better, and I'm glad I did. Simply put, Barb worked her ass off. She mastered every facet of her job and became a valued member of my crew, exceeding my expectations.

Dave Williams was a Canadian astronaut who worked as an ER doctor in his previous life. He is proud of his Welsh heritage and was the first person to broadcast from space in the Welsh language on his first space shuttle mission. Dave was completely unflappable.

Tracy Caldwell was flying her first mission. NASA selected Tracy when she was twenty-nine, right out of her Ph.D. program in chemistry. She looked young for her age, so she was treated as a bit of a kid by many of our astronaut colleagues, but her performance was top-notch. She was conscientious, incredibly detail oriented, and serious, but also fun to be around. Tracy turned thirty-eight on the sixth day of our mission.

Alvin Drew was assigned to the mission just three months before the flight. He flew helicopters in combat for the Air Force in their Special Operations Command and then went on to become a helicopter test pilot. He was not easily fazed and didn't seem thrown off by having been assigned to this flight so late, though it meant he would be constantly struggling to catch up.

For me, training to fly as commander was a completely new challenge. I had to learn my own role, as well as take responsibility for my crew—making sure everyone knew his or her job, recognizing each crew member's strengths and weaknesses, pulling us together as a team, and mentoring the rookies. Because we would have three first-time space flyers on our crew (Barb, Tracy, and Alvin), we were one of the least-experienced crews in shuttle history, with only four previous flights among the seven of us.

We went into quarantine in Houston ten days before launch, then flew to Florida and continued our quarantine there for the last four days. There is a NASA tradition, which some crews follow more closely than others, of pulling pranks on rookies. When the Astrovan pulled up to the launchpad, I said offhandedly to Tracy, Barb, and Alvin, "Hey, you guys remembered to bring your boarding passes, right?" They looked at one another quizzically as the four of us veterans pulled preprinted boarding passes out of our pockets.

"Don't tell me you didn't bring your boarding passes! They won't let you on the space shuttle without one!" I insisted. After an initial look of panic crossed their faces, the three rookies quickly caught on.

The closeout crew helped us get strapped into our seats, then climbed out of the shuttle and closed the hatch. Or tried to. The shuttle launch director announced that they couldn't tell whether our hatch was properly closed or not.

The shuttle hatch had presented problems before. The closeout crew, who knew the equipment better than anyone, felt it was closing properly, but no one wanted to risk our lives on that hunch. They shut the hatch, opened it again, shut it again, opened it again. We were all strapped tightly into our seats and couldn't see the hatch to give the closeout crew a visual confirmation. We were running out of time in our launch window.

Eventually Rick Mastracchio, who could stretch himself to see the hatch from his center seat on the flight deck, announced that the hatch was closed but that we now had an eighth crew member. One of the closeout crew had come into the space shuttle with us to inspect all the dogs—the bolts that attach the hatch to the surrounding structure—while the hatch was closed. He was able to confirm that it was working properly, and then the hatch was opened again so he could jump out. An ingenious and practical solution, and one I hadn't thought of myself.

THIS TIME I knew what to expect at launch, so I could enjoy it a little more, even looking out the window a bit. It had been nearly eight years since I last flew, and the sheer instantaneous power was still indescribable, the horizon pulling away faster than would seem possible. We reached orbit safely and, as on my previous mission, successfully got through the arduous job of turning the rocket ship into a spaceship.

Before I went to sleep, I got an email from the lead shuttle flight director telling me that nine pieces of foam had come off the external tank, three of which they thought had struck the thermal protection system on the bottom of the orbiter, similar to what had doomed *Columbia*—though in *Columbia*'s case, the damage was on the more critical reinforced carbon-carbon insulation on the leading edge of the wing. NASA didn't think it was a big deal—foam strikes could frequently be harmless—but were just letting me know out of an abundance of caution.

The next day, we conducted an inspection of the underside of the shuttle using a camera and laser scanners on the end of a boom attached to the robotic arm in an effort to pinpoint the damage. The images didn't reveal anything conclusive. We approached ISS the next day and flew the orbiter through a 360-pitch maneuver, a backflip to point the shuttle's heat shield so the station crew could capture up-close pictures. The photos showed an area of interest on a critical part of the belly of the orbiter near the right landing gear door; it was sizable enough that NASA decided to do a more focused inspection with the boom laser system after we docked. That inspection revealed a hole, about three inches by three inches, that went all the way through the silica thermal protection tiles down to the underlying felt.

As we scanned the area with the laser and looked at the images with the adjacent camera, my initial thought was, *Oh, shit!* The hole looked as though it went all the way through to the aluminum alloy that makes up the airframe. Later that evening, the ground emailed me photos of the damage. I printed out the most interesting pictures and carried them around in my pocket for the next couple of days.

There was a flurry of discussion on the ground about how this damage would affect our reentry. We didn't have a lot of options in this situation. We could try to fix the damage on a spacewalk by filling the hole with a special putty that had never been proven in flight, or take our chances and land as is. I talked over the options with my crew, mostly with Scorch, whose technical knowledge I held in particularly high regard. I also talked with our two spacewalkers, Rick and Dave, since they would have to do any repairs if we decided to go that route. We came to the conclusion that we could fix the damage if we had to, but we would trust the analysis done on the ground if they told us we could reenter safely. The press immediately wrote that the crew was in imminent danger.

Teams of experts on the ground were conducting analysis on the damage and how the heat of reentry would affect the tiles. They made a mock-up of the damaged tiles and put it in a testing facility where gases can be heated to very high temperatures and subjected to hypersonic speeds using a continuous electrical arc to simulate the effects of reentry. As I learned about the analysis they were doing, I had more and more confidence that the damage would not present a risk and that we should leave it as it was. Some NASA experts disagreed and thought we should do the repair. My concern was that one small bump from a crew member's tool or helmet could make the hole bigger, or create a new hole, and that the material and procedures for repairing tiles were still unproven. And, of course, any spacewalk presents inherent risks of its own.

THE DAY we were to return to Earth, we didn't dwell on the risk. We readied the orbiter and its systems, got suited up and strapped ourselves into our seats, and began the reentry process. As we slammed into the atmosphere and built up heat, we watched the hot plasma streaming past our windows and imagined the battering of the shuttle's heat shield. We all knew what could happen if our decision had been wrong.

"Passing through peak heating," Scorch said calmly. This was the point when *Columbia* had started to break up.

"Understand," I replied.

About twenty seconds later, we had passed the point where if the orbiter heat shield was burning through, we would have known about it.

"Looks like we dodged that bullet," I said. I couldn't help reflecting on our friends lost on *Columbia,* and I'm sure the rest of my crew was doing the same.

We were now inside the Earth's atmosphere, and as we slowed below the speed of sound, I took over the controls from the autopilot. I was flying the space shuttle for the very first time in Earth's atmosphere, and knew I would have only one chance to land.

As we dove seven times steeper than an airliner and descended twenty times faster, I felt the effects of gravity, vertigo, and a visual symptom called nystagmus, where your eyes jerk up and down. As we approached an altitude of two thousand feet, I tried to put these physical impairments out of my mind.

"Two thousand feet, preflare next," Scorch said.

"Roger preflare, arm the gear" was my response, acknowledging his call and asking him to arm the landing gear system. As we passed through two thousand feet, I started slowly and deliberately raising the orbiter's nose as I transitioned to a much shallower inner glide slope and started to rely more on the optical landing aids on the side of the runway and less on the orbiter's instruments.

At three hundred feet I told Scorch, "Gear down."

In response, Scorch pushed the button to lower the landing gear.

"Gear's down," he said.

From the time the landing gear were lowered until we landed was only about fifteen seconds. In that short period of time, I was trying to control the shuttle precisely in order to cross over the end of the runway at the correct height (twenty-six feet) and touch down at the correct speed (two hundred knots) with a rate of descent of less than two feet per second. We had a pretty heavy crosswind that day, which made all of this more challenging. I didn't touch down exactly on centerline, but by the time we came to a stop, I was perfectly in the center of the runway. I think most space shuttle commanders who were also carrier aviators—Navy pilots and Marines who have landed on a ship at night—would agree that landing the orbiter was easier, all things being equal, though still one of the hardest piloting tasks. What made it hard was doing a perfect landing when you've been in space and are tired, dizzy, and dehydrated. And of course, when the world was watching.

A FEW MONTHS AFTER I returned from STS-118, I was in D.C. to visit with members of Congress and went out to dinner with Mark's fiancée, Congresswoman Gabrielle Giffords. I'd first met Gabby in Arizona one afternoon a couple of years earlier when I went to pick up Mark at the airport. She was friendly, warm, and incredibly enthusiastic about her job as an Arizona state senator. I was impressed with her after our brief meeting, so much so that I joked to Mark that I wondered what she saw in him.

While we were eating, my phone rang, showing the number for Steve Lindsey, the chief of the Astronaut Office. As the fiancée of an astronaut, Gabby knew that when the chief astronaut calls at an unusual hour, you take the call.

"Scott, I'd like to assign you to a long-duration flight, Expedition Twenty-five and Twenty-six. You'd be the commander for Twenty-six."

I hesitated before speaking. It's always exciting to get a flight assignment, but spending five or six months on the International Space Station wasn't exactly what I had been hoping for.

"Honestly, I'd rather fly as a shuttle commander again," I said. "Is that possible?"

I knew the space shuttle inside and out, and I had only learned the basics about the Soyuz and the ISS. The Soyuz was a very different vehicle from the space shuttle, to say the least. I sometimes joked that Soyuz and the shuttle were similar in that they both carried people into space—and that was where the similarity ended. The Soyuz manuals and checklists were in Russian, for starters. And I would also need to learn more about the ISS, which had grown significantly in the last few years, inside and out.

I sighed. "When's the launch date?" I asked.

"October 2010."

"I understand. Let me talk to Leslie and my kids and I'll get back to you."

Five or six months away from home would be a long time, especially with Charlotte still so young. But I also knew I would take any flight assignment I was given. Leslie and the girls agreed this was an opportunity I couldn't pass up, and I said I would take it.

Among the things I had to do before I could turn my attention to this new assignment was follow up about my high PSA count. It wasn't alarmingly high, but it had jumped from its previous level, and the rate of change could be indicative of a problem. I visited a urologist, Dr. Brian Miles, at Houston's Methodist Hospital, who gave me two options: we could wait six months and see whether my PSA continued to increase, which would give us more information about whether I did have prostate cancer, and if so, how aggressive it might be. Or he could do a biopsy right then. I asked what the risk of a biopsy was.

"There's a low risk of infection at the biopsy site—that's really the only risk. People sometimes put it off as long as they can, though, because the procedure is uncomfortable."

"How uncomfortable?" I asked.

Dr. Miles paused while he thought about how to explain it. "Like small electric shocks through the wall of your rectum," he said.

"That sounds more than uncomfortable," I said, "but let's do it."

The procedure was as unpleasant as he said, but I didn't want to spend the next six months waiting to find out if I had cancer. If I did have it, I wanted to take care of it as soon as possible. Waiting could jeopardize my chances to fly my next mission or put the ISS schedule at risk.

A few days later, I learned I had a relatively aggressive strain of prostate cancer. Some types are so slow growing that men can live with them for decades and not be affected. The type I had would not create any adverse effects for a while, but if left untreated it would likely kill me in twenty years or so (I was forty-three).

When you are told you have cancer, especially an aggressive one, your mind immediately runs wild. Is this pain in my arm a metastasized tumor? Is the cancer spreading to my brain? I think this is a normal reaction to have, even for people who have access to top-notch care. I was immediately sent for a full-body CAT scan, and there was no indication the cancer had spread, which did a lot to set my mind at ease.

One of the first people I spoke to was my crewmate Dave Williams, who had had surgery for prostate cancer himself. As a doctor, he was able to offer good advice. He went to several meetings with me to talk to the surgeon about treatment options, along with the NASA flight surgeons.

Meanwhile, I called my brother and told him to get himself checked out. Since we were identical twins, we had a nearly identical genetic blueprint and therefore similar risks. When Mark got checked, it turned out he had the same type of prostate cancer.

I decided on a robotic radical retropubic prostatectomy, a surgery that would remove the entire prostate and leave me with a daunting recovery. It also brought with it a risk of bad outcomes like impotence or incontinence. There were less aggressive options—radiation therapy

My phony juggling act: the fruit had been sent to the ISS on the Japanese HTV resupply vehicle.

Quiet time in the Cupola

Samantha storing her science samples in one of three −98°F freezers in the Japanese module

One of the incredible auroras we sometimes flew through

The most powerful hurricane ever recorded, Patricia, as it approached the western coast of Mexico, October 23, 2015

Kjell in the U.S. laboratory module demonstrating the joys of living in zero gravity

One of my favorite places on Earth, New York City

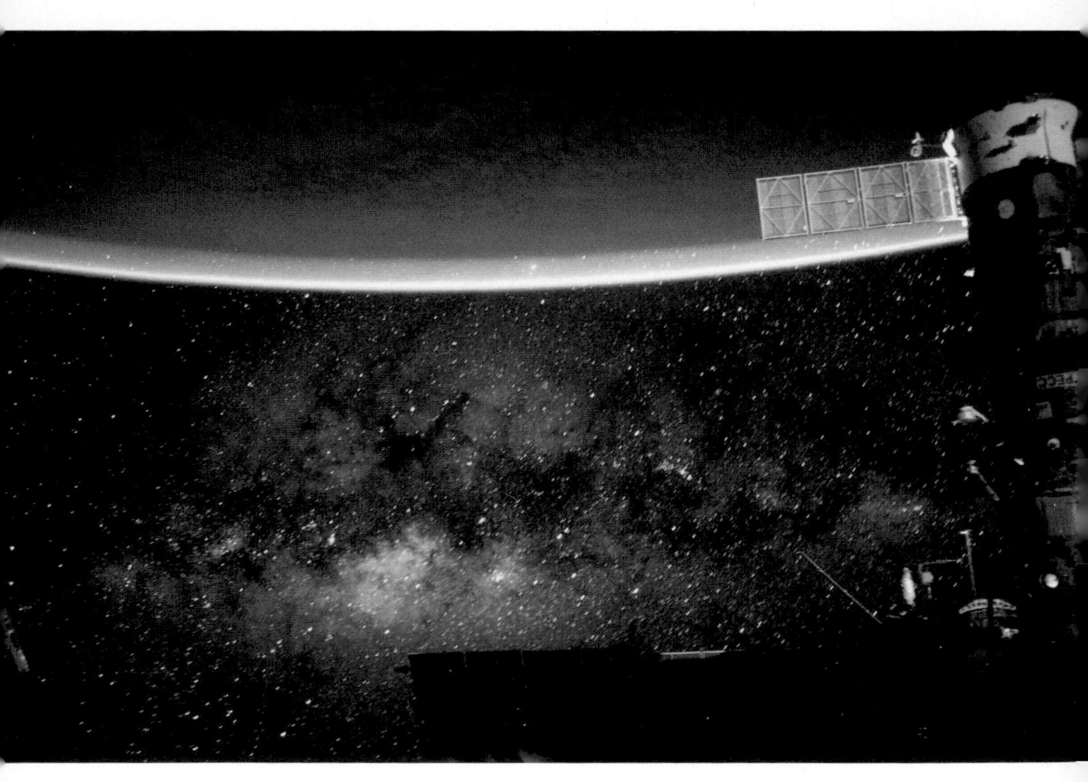

One of my favorite places to admire off Earth, our Milky Way galaxy

Pollution from the Indian subcontinent as it butts up against the Himalayas

Sunrise aboard the ISS

After the parachute opens we drift peacefully down to Earth, our time on the ISS over.

Our Soyuz spacecraft as it crashes back onto Earth on March 2, 2016

I'm being helped out of the capsule but am feeling triumphant.

Mark and me after my year in space. Not bad for two kids from New Jersey.

or a combination of less drastic surgery and radiation. It could take up to two years to determine whether radiation had successfully eliminated the cancer, though, and I didn't want to have to wait that long to fly again. More important, because astronauts are exposed to radiation in space, our flight surgeons keep track of a lifetime radiation limit for each of us. I didn't want to run up my lifetime limit if I could possibly avoid it. The robotic surgery was the option most likely to wipe out the cancer for good and minimize the risks to my career.

I underwent surgery in November 2007. My recovery took a long time, just as my surgeon had said it would, and it wasn't fun. I had a urinary catheter for a week and a drain for lymphatic fluid in my side for weeks. One of the flight surgeons stopped by my home to check on my recovery one evening, and he decided the drain catheter was ready to come out. Standing in my living room, he simply yanked it out with all his might, and without much warning. I had no idea the thing was three feet long until I saw, and felt, it being ripped out of my body. I felt like I was William Wallace getting eviscerated in the movie *Braveheart.*

Despite the long recovery overall, I was aggressive in getting my NASA qualifications back, and I was able to start flying again in January. Getting back into the pool where we do spacewalk training, however, took much longer, as there was concern that the crotch of the spacesuit would put pressure on the area that was still healing. Due to the skill of Dr. Miles and the NASA flight surgeons, I returned to near normal in due time. The following year, I was in the operating room while Dr. Miles performed Mark's operation, and I got to hold Mark's prostate in my gloved hand before it was sent to pathology. The tumor was on the opposite side from mine, a mirror image, just like the opposite birthmarks on our foreheads.

IN EARLY 2008, I started training for my mission to the space station in earnest. I would be launching with two Russians, Sasha Kaleri and Oleg Skripochka, and when we got to orbit we'd be joining Shannon Walker, Doug Wheelock, and Fyodor Yurchikhin. Three months into my mission, Shannon, Doug, and Fyodor would return home and be replaced by Cady Coleman, Italian astronaut Paolo Nespoli, and Dima Kondratyev. I traveled often to Russia, Japan, and Germany to train with their respective space agencies.

I had a lot of experience working with the Russians in Star City by this point, which was good, because it would lighten my training load some, but I would still spend a significant amount of my time there. I'd learned a great deal about the differences between our cultures—that Russian people's behavior toward strangers is indifferent to the point of seeming cold, which to Americans can feel rude, but that once I got to know individual Russians well, their behavior toward me was warm and affable. My friendships with people there reached a depth that would take many Americans years to reach.

The instructors we had at the European Astronaut Centre in Cologne, Germany, came from countries throughout Europe. I found it enriching to work with such a diverse group of people, but the training culture itself was purely German—precise almost to a fault. This could be a bit maddening at times to someone like me, who doesn't care how the sausage is made. I prefer to be told what needs to be done and how to do it, leaving the nuanced details for the ground to worry about. I spent four weeks training in Cologne. I was enamored of the architecture, particularly the High Cathedral of Saint Peter, a thirteenth-century colossus that stands proudly on the banks of the Rhine.

Compared to my Russian and European colleagues, the Japanese people I met were much more outwardly polite and deferential to strangers, but it took much longer to get past that polite stage and get to something more familiar. Because my Japanese colleagues were polite to everyone, I found it hard to tell whether I had established good working relationships with them. This concerned me because I knew my directness could often be taken the wrong way and for some be off-putting.

To train with the Japanese space agency, I traveled to Tsukuba, a city of about 200,000 located about fifty miles northeast of Tokyo. I was joined there by my future crewmate Doug Wheelock and by Tracy

Caldwell from my STS-118 crew, whom I was now backing up prior to my own expedition. One evening as we walked to one of the nearby restaurants, we passed a dessert truck with an English inscription on one side. In large letters, the truck was labeled "Marchen & Happy for You," under which was printed an odd sort of prose poem:

Beginning was from only one car.
Fumipasu Hasebe who is the founder who was 23 years old
A little brought the method of the vehicle sale completely different from former close to a grope and completion in 1998.
Creation of the scrupulous interior and jokespace.
The way a visitor can share not only the taste but a style.
It is thought of wanting to provide many people with a joke style with a delicious dessert.
It is the language and "Marchen & Happy for you,"
Which do not change now, either but are hung up.
It does not still change at all. And present
The top in the field of move sale in aimed at and it is under business in various parts of national.
It is our ship also to your town.

The first time we saw this truck, we stopped and read the "poem" out loud, fascinated by its almost-sensical approximation of colloquial English. The Marchen & Happy for You truck became a landmark for English speakers visiting Tsukuba, and we always made sure to point it out to newcomers, watching while they read it and tried to make sense of it. I took a picture of it with my phone one day and showed it to one of the instructors at the Tsukuba Space Center who had strong English skills. I read him the text and asked him, "Does this make any sense to you?"

"Of course," he answered. "What don't you understand?"

This only increased our fascination with the ice cream truck and the linguistic and cultural differences it symbolized. To this day, many years later, when my former colleagues and I get together and talk about old times, especially training in other countries, sooner or later someone will bring up a picture on his or her phone and start reading out loud from it: "Beginning was from only one car. . . ." Laughter inevitably ensues, sometimes followed by crying from laughing so hard. The dessert truck poem reminds us of the "lost in translation" aspect of being Americans in Japan, but it also reminds us of that intense period of training for an ISS expedition, and the way these shared experiences brought us together.

As WITH most marriages that begin with the groom thinking about how to get out of the ceremony, my marriage to Leslie was not a happy one. Leslie was a good mother, and she continued to take care of things on the home front so I was free to work demanding hours at NASA, including frequent travel. After Samantha was born, I periodically tried to initiate conversations about our relationship and the possibility of ending our marriage. These conversations never went well. Our talks always ended with Leslie saying that if I ever tried to leave her, she would destroy my career and I would never see my child again. I was shocked and saddened that she would threaten me this way, but I also understood that emotions were running high and we all had a lot at stake.

We decided to try counseling. I was reluctant at first because I thought it might affect my chances to fly in space. I had been asked in the process of interviewing with NASA whether I had ever sought counseling or psychiatric help, and, having truthfully said no, I didn't want that to change. Astronauts never knew exactly why we got flight assignments or what kept us from getting them, so the instinct to avoid negative attention or controversy was deeply ingrained. But I agreed because I thought it might help, and Leslie wanted to try. The day of our first appointment, we were waiting in the reception area when the door to the counselor's office opened and out came a senior management astronaut and his wife, both of them wearing the stony expressions of people who have been through an emotional wringer.

He and I silently acknowledged each other, and although I wondered whether having been seen there would have consequences, I at least knew it wasn't unheard of for an astronaut to seek help for a troubled marriage.

The counselor wasn't able to help us much, and our marriage continued to deteriorate. Meanwhile, I dropped the subject of our marriage each time Leslie threatened me. After Charlotte was born, and "child" changed to "children," the stakes were even higher. So we settled into a semi-friendly arrangement in which she took care of our children and home and I pursued my career. I was gone a lot, which minimized the opportunities for tension and fighting, and we both liked entertaining and having people around, so even when I was home there wasn't much chance for serious drama. We continued this way for years.

In the spring of 2009, I was back in Japan. I'd been looking forward to the trip, but once I was there, I felt crappy and the weather was gray and dull. I had a bad cold, was exhausted from jet lag, and was in a foul mood the whole time. I dragged myself through classes and training sessions all day and then collapsed in my tiny economy hotel room at night. It was then that I realized that despite being unhappy in Tsukuba, I didn't want to go home to Leslie. I would rather be on a business trip feeling miserable than in my own home.

I visited my grandmother the day after I returned to the States. My father's mother, Helen, whose home had been such a sanctuary for Mark and me when we were boys, was now in her nineties and living in a nursing home in Houston. She had taken a turn for the worse, and while I sat with her, holding her fragile hand, I thought about what a comforting presence she had been when we were little, when she took us to the botanical gardens and sang us to sleep. That had been forty years ago, and now age had robbed her of her vitality. Where would I be when I was her age, decades in the future? If I was lucky enough to be alive still, what sort of life would I have to look back on? How was I going to spend the rest of my time on Earth?

The very next day, I called Leslie from work and let her know I would be coming home early and that I needed to talk to her alone when I got there. At home, I told her I would always respect her as the mother of our children, and I would always take care of my daughters, but I wanted a divorce.

As I had anticipated, she repeated her threats and reminded me she had evidence that I had been unfaithful.

"I can understand that you're angry," I said, "but this is what I've decided. I hope you can move on. But do what you need to do."

I'd hoped to have an amicable split, for the benefit of our daughters. Samantha was now fourteen, an especially vulnerable age to deal with this kind of family upheaval, and Charlotte was five. I thought it was important to show the girls that adults could work through their problems calmly, cooperatively, with generosity of spirit, and with an emphasis on the well-being of the children. This was not to be.

When Samantha and Charlotte got home from school, I gathered myself and spoke to them as calmly as I could, trying to make things seem cordial and positive, though they could tell from their mother's face that this was nothing of the sort. Samantha was more upset than Charlotte—she was old enough to understand what a big change this was going to be. I tried to assure her that I would do everything I could to keep her life stable. Charlotte didn't seem very interested in the conversation and spent the whole time playing with a rubber band—wrapping it around her wrist, unwrapping it from her wrist, her bangs hiding her eyes. After a while, Leslie asked her if she had any questions.

Charlotte's round little face tipped up at me. She met my eyes, and I tried to read her expression. Then she held out the rubber band to me and asked, simply, "Is this your rubber band?"

This gesture was typical of Charlotte. She was trying to change the subject away from the topic that was causing everyone so much pain, and at a moment when I was so concerned about my daughters and how their world was about to be blown apart, *she* was trying to give *me* something.

When I put my head on the pillow that night, I felt more at peace than I had in months, maybe years. Maybe I would never fly in space again, but I was going to try to live a life I wouldn't regret when I was old.

Leslie carried out one of her threats by moving away with the children, but in the end our divorce didn't affect my career as I'd feared it might. She is still angry at me for ending our marriage. Yet when I started seeing Amiko, Leslie was surprisingly warm toward her. Whatever animosity she continued to feel toward me, she didn't extend to Amiko, which a lot of people might have done in her situation.

Not long ago, Leslie and Amiko were consulting on the phone about some travel arrangements for Charlotte, when Leslie said to her, "I want you to know that you have always been great to co-parent with. My girls just love you, and that makes me love you too." Amiko hung up the phone with tears in her eyes. She has been through a lot with my family, and these kind words meant the world to her. I know some people who, after going through a difficult divorce, say they wish they had never married their spouse or had never even met him or her. I can honestly say I have never felt this way. Leslie has been an important part of my life, and though I wish we could be on better terms, I have never regretted my decision to marry her, and I am eternally grateful for Samantha and Charlotte.

15

October 28, 2015

Dreamed Kjell and I were going to go skydiving together. We went up in a plane, and as I was standing near the doorway Kjell jumped out without his parachute. I watched his face change as he realized his mistake, a look of horror overtaking him as he slowly fell away from me. I didn't have my own parachute on yet, so I was scrambling around looking for one so I could jump out and catch Kjell before he could hit the ground. I searched frantically through piles of junk in the plane. After a while I knew it must be too late, but I kept looking anyway until I woke up.

I'M FLOATING in the U.S. airlock, wearing a 250-pound spacesuit, while the air is slowly pumped out. I can't see Kjell's face because we are crammed into a space the size of a compact car, at odd angles, his head down near my feet. I've been in the suit for four hours now. Kjell is wearing the only extra-large spacesuit on station because he couldn't fit into the large-sized one, so I'm wearing a suit that's clearly too small for me, feeling like ten pounds of potatoes stuffed into a five-pound bag. I'm already tired and sore.

"How you doing, Kjell?" I ask, staring directly at his boots.

"Great," Kjell says and gives a quick thumbs-up I can barely see through the bottom of my visor. Any normal person, upon experiencing the air leaving the airlock around him, would be somewhere on the scale between apprehensive and terrified. But Kjell and I have trained for this, our first spacewalk, for a long time, and we feel prepared and confident in the equipment and the people who are keeping us safe.

Suddenly a series of loud bangs reverberates through the airlock, a sound I've never heard in training. It's like someone knocking on a door loudly and urgently. Then it's quiet. Has something gone wrong? Should we be doing something? I mention the sound to the ground, and they tell me that it's normal, one of the things that happens when the air is sucked out of the airlock. No one thought to tell us about it in training, or maybe they just forgot to mention it, or maybe they did and I forgot. I've practiced this moment many times, wearing a spacesuit and being lowered into a giant swimming pool containing an ISS mock-up at JSC, but it's different doing it for real, in space, with no safety divers to help us out if things go wrong.

Once the airlock is nearly at vacuum, Kjell and I do a series of checks on our spacesuits to make sure they are not leaking. This process consists of a series of switch throws and slides of a lever, all of which are extremely difficult to do while wearing the suit's gloves, sort of like trying to change a car's tires while wearing a baseball glove. To make things worse, we can't see the controls, so we have to use mirrors attached to our wrists to see what we are doing (the labels on the controls are written backward so we can read them).

Looking ahead at the procedures, I see that once the airlock is down to a complete vacuum, each of us will turn our water switch on, which will allow water to flow through the cooling system to control the temperature in our suits. We can't do this prematurely because the water can then freeze and crack the lines. As the air continues to escape the airlock, I consider warning Kjell that the water switch is easy to flip accidentally. It's right next to a similar-looking switch that we use often to silence alarms or scroll through lines of status messages on a small LCD screen. But I tell myself that Kjell is as well trained as I am for this spacewalk. I'm not going to micromanage him.

When the airlock is not quite at a vacuum, Kjell says, "Houston—and Scott—I just hit my water switch on/off."

Shit! I think, but don't say. I take a breath to steady myself. "You cycled it?" I ask. He's just done the very thing I decided against warning him about.

"Yeah."

Our capcom for the spacewalk is Tracy Caldwell Dyson, my crewmate from my second shuttle flight—she gained a new last name through marriage in the intervening time. "Houston copies," Tracy responds. "Kjell, can you tell us how long it was on?"

"Less than half a second," Kjell says. He sounds dejected. We've already spent hours today—and entire working days over the past two weeks—getting ready for this spacewalk. We do not want to have to start all over, not to mention the possibility of damaging the $12 million suit.

While spacesuit experts on Earth confer about how to proceed, I'm pissed at myself for not warning Kjell. Knowing the way NASA works, we are aware there is a very good chance they will not allow us to continue. If that happens, it will be because the experts cannot guarantee Kjell's safety, and the most important thing is that we both finish the day alive. On the off chance NASA will let us continue, I need Kjell to keep his head in the game.

"It's happened before, Kjell," I tell him. "It'll happen again."

"Yeah," Kjell answers, sounding dispirited.

"Don't worry about it," I say, wishing I could make eye contact to see how he's doing.

"No worries," Kjell replies in a flat tone completely at odds with his words. Astronauts have seen their careers permanently affected by mistakes like this.

"It'll be all right," I say, talking to myself as much as I'm talking to him now.

Suit experts on the ground are still discussing whether to proceed, and what precautions we'll need to take. Meanwhile, we are told we can open the hatch and enjoy the view while they decide on a course of action. As I put my hand on the handle, I realize that I have no idea whether it will be day or night outside. I unlock the hatch handle and crank it, releasing the "dogs." Now I have to simultaneously translate the hatch toward my chest and rotate it toward my head, which is challenging, because with nothing to hook my feet onto, I'm pulling myself toward the hatch almost as much as I'm pulling it toward me.

I tug and push and pull for a few minutes, and finally the hatch cracks open. The reflected light of Earth rushes in with the most abrupt and shocking clarity and brightness I've ever seen. On Earth, we look at everything through the filter of the atmosphere, which dulls the light, but here, in the emptiness of space, the sun's light is white-hot and brilliant. The bright sunshine bouncing off the Earth is overwhelming. I've just gone from grunting in annoyance at a piece of machinery to staring in awe at the most beautiful view I've ever seen.

Inside my spacesuit, I feel like I'm in a tiny spacecraft rather than wearing something. My upper body floats inside the hard torso, my head encased in the helmet. I hear the comforting humming noise of the fan moving the air around inside my suit. The helmet has a faint chemical smell, not unpleasant, perhaps the anti-fog solution our visors are treated with. Through the earpiece built into my comm cap, I can hear the voices of Tracy in Houston and Kjell just a few feet away from me out here in outer space—that, and the strangely amplified sound of my own breathing.

The surface of the planet is 250 miles below my face and whizzing by at 17,500 miles per hour. It takes the ground about ten minutes to tell Kjell and me to go outside the hatch, where we can move around better, so I can check over Kjell's suit for a leak. In the cold of space, a leak would look like snow shooting out of the backpack of the suit. If I don't see snowflakes, we may be allowed to continue.

I grab both the handrails on either side of my head, getting ready to pull myself out. The airlock's hatch faces the Earth, which would seem to be the direction we would call "down." When we trained in the pool, the hatch was facing toward the floor, which always felt like down. Though I was neutrally buoyant in the pool, gravity still forced me toward the center of the Earth, providing a clear sense of up and down. For the hundreds of hours we practiced for this spacewalk, I got used to the idea of this configuration.

Once I'm about halfway through the hatch, though, I have a transition in perspective. Suddenly I have the sensation of climbing up, as if out of the sunroof of a car. The large blue dome of Earth hovers over my head like some nearby alien planet in a sci-fi film, looking as if it could come crashing down upon us. For a moment, I'm disoriented. I'm thinking about where to look for the attachment point, a small ring where I will hook my safety tether, but I don't know where to look for it.

Like any highly trained pilot, I know how to compartmentalize, to push thoughts out of my mind that aren't helping me to complete the task at hand. I focus on what is immediately in front of me—my gloves, the handrail, the small labels on the outside of the station I've familiarized myself with through countless hours of training—and ignore the looming Earth above and the feeling of disorientation it creates. I don't have time for it, so I set it aside and get to work. I take the hook from my safety tether off my mini-workstation, a high-tech toolholder attached to the front of my spacesuit, and secure it to one of the rings just outside the airlock, checking to make sure the hook is in fact closed and locked with complete certainty. Like putting an airplane's landing gear down before landing, this is one of those things you absolutely do not want to screw up.

During my last long-duration mission on ISS, two of the Russian cosmonauts, Oleg Skripochka and Fyodor Yurchikhin, did a spacewalk together to install some new equipment on the outside of the Russian service module. When the two of them came back inside, they both looked shaken, Oleg especially. I assumed at first his reaction was to being outside for the first time, and it wasn't until this yearlong mission that I learned all the details of what had happened that day: during their spacewalk, Oleg had become untethered from the station and started to float away. The only thing that saved him was hitting an antenna, sending him tumbling back toward the station close enough to grab on to a handrail, saving his life. I've often pondered what we would have done if we'd known he was drifting irretrievably away from the station. It probably would have been possible to tie his family into the comm system in his spacesuit so they could say good-bye before the rising CO_2 or oxygen deprivation caused him to lose consciousness—not something I wanted to spend a lot of time thinking about as my own spacewalk was approaching.

The U.S. spacesuits include simple propulsive jets we could use to maneuver ourselves in space in case our tether breaks or we screw up, but we would not want to rely on them or, truth be told, try them out at all. The only way we practiced using the jet packs during our training was with virtual reality simulations, during which astronauts sometimes ended up running out of fuel or missing the space station altogether. I'm acutely aware that if I become detached and run out of fuel and the station is just one inch from my glove tips, it may as well be a mile. The result will be the same: I will die.

Once I'm certain my tether is secured, I remove Kjell's tether from me and attach it to the outside of the station as well, being just as careful to double-check it as I was with my own. Kjell starts handing me bags of equipment we will need for our work, and I secure each of them to the circular handrail outside the airlock. Once we have everything we need, I give Kjell the go to exit. The first thing we do once we are both outside is our "buddy checks," looking over each other's suits from head to toe making sure everything is in order. Tracy talks us through it from mission control, telling me step-by-step how to check Kjell's PLSS (portable life support system, the "backpack" we wear with the spacesuit) for signs of water having frozen in the sublimator. It looks completely normal—there are no snowflakes, I'm happy to report to the ground. Kjell and I both breathe a sigh of relief. Our spacewalk will proceed. (Later we would learn that some of the engineers wanted to call off the spacewalk and the lead flight director overruled them.) We go over each other's helmet lights, helmet cameras, mini-workstations, jet pack handles, checking to make sure everything is properly stowed. One of Kjell's jet pack handles is not—it was partially deployed while Kjell was on his way out of the airlock—and one of mine was as well. After fixing them, we check our tethers one more time. You can't be too careful with tethers. Nearly five hours after getting into our spacesuits, we are ready to get to work.

. . .

FOR NEARLY as long as human beings have been going into space, we have been determined to climb out of the spacecraft. It's partly just to achieve the fantasy of a human being floating alone in the immensity of the cosmos, nothing but an umbilical connecting him or her to the mother ship. But spacewalks are also a practical necessity for exploration. The ability to move from one spacecraft to another, to explore the surfaces of planetary bodies, or (especially relevant to the International Space Station) to perform maintenance, repairs, or assembly on the exterior of the spacecraft—all are crucial to long-term space travel.

The first spacewalk was carried out in 1965 by cosmonaut Alexei Arkhipovich Leonov. He opened the hatch of his Voskhod spacecraft, floated out on an umbilical, and reported to Moscow: "The Earth is absolutely round"—probably to the dismay of flat-earthers everywhere. It was a triumphant moment for the Soviet space program, but after twelve minutes, Alexei Arkhipovich found that he could not get back through the hatch. Due to a malfunction or poor design, his spacesuit had inflated to the point that he could no longer fit through the narrow opening; he was forced to let some of the precious air out of his suit in order to struggle back through. Doing so caused the pressure to drop so much he nearly passed out. This was not an auspicious beginning to the history of spacewalks, but since then more than two hundred people have successfully suited up to float out an airlock into the blackness of space.

While some of the challenges of spacewalks have gotten easier, they are no less dangerous. Just a few years ago, astronaut Luca Parmitano's helmet began filling with water while he was outside, raising the terrifying specter of an astronaut drowning in space. Spacewalks are much riskier than any other part of our time in orbit—there are so many more variables, so many pieces of equipment that can fail and procedures that can go wrong. We are so vulnerable out there.

As pilot and commander of the space shuttle, I never had the chance to do a spacewalk. The mission specialists went through the hundreds of hours of training necessary to work outside the spacecraft while I trained to fly and command the mission. For most of the shuttle era, those of us designated as pilots knew that, because of this division of labor, we would never have the chance to put on a spacesuit and float out into the cosmos. A shuttle could return safely with a missing or injured mission specialist, but a missing pilot or commander would be much more problematic. But we are now in another era of spaceflight, and this mission on ISS has given me the chance.

Getting ready to go outside takes a great deal of time. We plan in advance as thoroughly as we can what we will do and in what order, to minimize problems and to maximize efficiency and performance. We prepare the suits, check and double-check all of the components that will keep us alive in the vacuum of space, and organize and prepare the tools we will use—custom-designed for use in zero gravity with our bulky gloves.

I've been up since five-thirty this morning and have been hurrying to stay ahead of the timeline all day. I got into a diaper and the liquid cooling garment we wear under the spacesuits, like long underwear with built-in air-conditioning once it's connected to the suit. Then I ate a quick breakfast I'd laid out the night before to save time and made my way to the airlock to start getting suited up. My goal was to be out of the airlock early—my philosophy is that for complicated jobs, if you aren't ahead of schedule, you're already behind.

Kjell and I spent an hour breathing pure oxygen to reduce the amount of nitrogen in our blood so we wouldn't get the bends (decompression sickness). Kimiya is the intravehicular crew member (IV) for this spacewalk, responsible for helping us get dressed, managing the procedure for prebreathing oxygen, and controlling the airlock and its systems. His tasks might seem mundane, running down a checklist with hundreds of steps, but his job is critical for Kjell and me. It's practically impossible to get in and out of a spacesuit without help, and if Kimiya makes even the smallest mistake—puts on my boot incorrectly, for instance—I could die a horrible death. My suit includes a life support system that keeps oxygen flowing, scrubs the carbon dioxide that I exhale, and keeps cool water flowing through the tubes covering my body so I don't get overheated. Although weightless, the suit still has mass. It's also stiff and bulky, making it difficult to maneuver.

I slid into the pants of the suit, and Kimiya helped me squeeze into the hard upper torso. Nearly dislocating my shoulders and hyper-extending my elbows, I pushed my arms into the sleeves and my head through the neck ring. Kimiya connected my liquid cooling garment umbilical, then sealed the pants to the torso. Each connection between pieces of the suit is critical. The last step was to put my helmet on. My visor had been fitted with Fresnel lenses to correct my vision without me having to wear glasses or contacts. Glasses can slip, especially when I'm exerting myself and sweating, and I can't adjust them when I'm wearing my helmet. Contact lenses would be an option, too, but they don't agree with my eyes.

Once we were suited up, Kimiya floated us into the airlock—first me, then Kjell—allowing us to conserve our energy for what was to come. We floated and waited for the air to be pumped out of the airlock and back into the station. Air is a precious resource, so we don't like to vent it out into space.

Tracy's voice breaks the silence: "All right, guys, with Scott leading, we will begin translating out to your respective work sites."

By "translate," she means to move ourselves, hand over hand, along a path of rails attached to the outside of the station. On Earth, walking is done with the feet; in space, especially outside the station, it's done with the hands. This is one of the reasons why the gloves of our space-suits are so critical.

"Roger that," I tell Tracy.

I translate out to my first work site, on the right side of the giant truss of the space station, occasionally looking back to see how my tether is routed and making sure it doesn't get snagged on anything. At first, I feel like I'm crawling hand over hand across a floor. I'm immediately struck by how damaged the outside of the station is. Micrometeoroids and orbital debris have been striking it for fifteen years, creating small pits and scrapes as well as holes that completely penetrate the handrails, creating jagged edges. It's a little alarming—especially when

I'm out here with nothing but a few layers of spacesuit between me and the next strike.

Being outside is clearly an unnatural act. I'm not scared, which I guess is a testament to our training and to my ability to compartmentalize. If I were to take a moment to ponder what I'm doing, I might completely freak out. When the sun is out, I can feel its intense heat. When it sets, forty-five minutes later, I can feel the depths of the cold, from plus to minus 270 degrees Fahrenheit in minutes. We have glove heaters to keep our fingers from freezing but nothing for our toes. (Luckily, my ingrown toenail has healed after a few weeks without any further intervention or this would be even more uncomfortable.)

The color and brilliance of the planet, sprawling out in every direction, are startling. I've seen the Earth from spacecraft windows countless times now, but the difference between seeing the planet from inside a spacecraft, through multiple layers of bulletproof glass, and seeing it from out here is like the difference between seeing a mountain from a car window and climbing the peak. My face is almost pressed against the thin layer of my clear plastic visor, my peripheral vision seemingly expanding out in every direction. I take in the stunning blue, the texture of the clouds, the varied landscapes of the planet, the glowing atmosphere edging on the horizon, a delicate sliver that makes all life on Earth possible. There is nothing but the black vacuum of the cosmos beyond. I want to say something about it to Kjell, but nothing I can think of sounds right.

My first task is to remove insulation from a main bus switching unit, a giant circuit breaker that distributes power from the solar arrays to the downstream equipment, so the unit can later be removed by the main robotic arm. This is a job that would normally require a spacewalk, but we are trying to use the robot arms to do more work.

Kjell's first task is to put a thermal blanket on the Alpha Magnetic Spectrometer, a particle physics experiment. It's been sending back the kind of data that could alter our understanding of the universe, but it needs to be protected from the sun if it's going to continue doing its job—it's getting too hot. The spectrometer was delivered to the station by the last flight of *Endeavour* in 2011, which was commanded by my brother. Neither of us would have guessed five years ago that I would be leading a spacewalk to extend its lifespan.

The Hubble Space Telescope and other instruments like the AMS have transformed our understanding of the universe in recent years. We had always assumed that the stars and other matter we could observe—200 billion galaxies each with 100 billion stars on average—made up all of the matter that existed. But we now know that less than 5 percent of the matter in the universe is actually observable. Finding dark energy and dark matter (the rest of what's out there) is the next challenge for astrophysics, and the AMS is searching for them.

Removing and stowing the insulation from the main bus unit is a relatively simple task for a spacewalk, but as with everything we do in zero g, it is harder than you would think—sort of like trying to pack your suitcase if it were nailed to the ceiling. The focus required to do even simple work in space is daunting, similar to the focus required to land an F-14 Tomcat on an aircraft carrier, or land the space shuttle. But in this case I have to maintain that focus all day, rather than only for a matter of minutes.

The three most important things to keep track of today are what I think of as the three T's: tethers, task, and timeline. From moment to moment, I have to be aware of my tethers and whether they are properly attached. There is nothing more important to my continued survival. In the medium term, I have to focus on the task at hand and on completing it properly. And in the long term, I have to think about the overall timeline for the spacewalk—the scheduled sequence of tasks planned out to make the best use of our finite suit resources and our own energy.

When I finish removing the insulation and stuffing it into a bag, I get congratulations from the ground for a job well done. For the first time in hours, I take a deep breath, stretch as best I can in the stiff spacesuit, and look around. This would normally be a good time to break for lunch, but that's not on today's schedule. I can sip water through a straw in my helmet, but that's it. I'm making good time and still have a lot of energy. *We are going to be able to nail this spacewalk,*

I think to myself. As the day goes on, it will become clear that this is a false sense of confidence.

The next task for me is working on the end effector, the "hand" of the robot arm. Without it, we can't capture and bring in the visiting vehicles that deliver food and other necessities to the U.S. side of the station. Once I'm secured in a foot restraint, I realize how lucky I am: rather than facing the blank exterior of an ISS module, as spacewalkers usually do (and as Kjell is at this moment), I am facing out toward the Earth. I can watch the stunning view splayed out below my feet as the Earth goes by while I work, rather than having to turn around and look out the corner of my eye during the rare free moment. I feel like Leonardo DiCaprio at the bow of the *Titanic*, and I'm king of the world.

While I was training for this mission, I practiced greasing a replica of this end effector, using tools identical to the ones I'm using up here. While I practiced, I wore a duplicate of my spacesuit gloves. But the experience is still disorientingly different now that I, the grease gun, and the grease are all floating in space, the sun rising and setting spectacularly every ninety minutes, the planet spinning majestically underfoot. The grease gun is well designed, like a high-grade version of a caulking gun from a hardware store, but it's awkward to use with the fat-fingered gloves of the pressurized suit. For several hours, I wield this cumbersome tool like a five-year-old with finger paint. The grease goes everywhere. Small beads of grease jump off the gun as if they have a will of their own to explore the cosmos. Some of them come right toward me, which could pose a serious problem; if grease starts to coat the faceplate of my helmet, I may not be able to see to find my way back in. This task is taking much longer than had been scheduled, and soon my hands are aching to the point where I start to think I might not be able to move them. Of all the things that are tiring about this spacewalk, the amount of effort it takes to manipulate the gloves is by far the worst. My knuckles are rubbed raw, the muscles beyond fatigued, and I still have a great deal left to do. I work with Kimiya as he precisely maneuvers the robot arm to place it exactly where I need it. I put grease on the end of a long wire tool and stick it into the dark hole of the end effector. I can't see in there and can only hope the grease is going in the right place as I blindly feel around.

This task is taking so long I know I won't get to complete some of our other scheduled activities. Kjell is running long as well; the cables he is routing to enable future visiting vehicles to dock are proving to be as difficult to wrangle as my grease gun. We are well past the six-and-a-half-hour mark when we start getting organized to call it a day and head back to the airlock. Despite having consumables that would last another few hours, we have to leave enough time to deal with any unexpected problems.

We still have the toughest part of the spacewalk in front of us: Kjell and I must maneuver ourselves back into the airlock. Kjell goes first and guides his bulky suit through the opening without getting hung up on anything. Once inside, he attaches his waist tether. Then I release his safety tether, which is still connected to the outside of the station, and attach it to myself, then release my own. I swing my legs over my head and flip upside down into the airlock, so I will be facing the hatch to close it.

By the time we are both inside we are breathing hard. Closing the hatch, absolutely mandatory, will be much harder than opening it, with the fatigue from the spacewalk taking its toll. My hands are completely spent.

The first step is to close the thermal cover of the outside hatch, which has been severely damaged by the sun, like most of the equipment exposed to its harsh rays. The cover doesn't fit right anymore—it's assumed the shape of a potato chip—and it takes a lot of finesse to get it secured properly. With the thermal cover closed, it's time to get hooked back up to the umbilical that provides oxygen, water, and power to the suits via the station's systems rather than the suit itself. This isn't an easy task either, but after a few minutes we manage to get them connected properly.

Despite my fatigue, I manage to get the hatch securely closed and locked. As the air hisses in around us, Kjell and I are still breathing hard from the work of getting back inside. We will have a wait of about fifteen minutes, punctuated by a few leak checks, to make sure the hatch is properly closed while the airlock returns to the pressure of the station. As we wait, I struggle to equalize my ears by pressing my nose against a pad built into my helmet and blowing (this Valsalva device is designed to replicate the effect of holding our noses). This requires much more force than I thought it would, and later I will discover I have burst some blood vessels in my eyes in the process.

We have been in these suits for eleven hours now.

At some point during the repressurization process, we lose comm with the ground. We know it means that for at least a while we aren't being broadcast on NASA TV and can say what we like.

"That was fucking insane!" I say.

"Yeah," agrees Kjell. "I'm beat."

We both know we will have to do another spacewalk in nine days.

When the hatch opens and we see Kimiya's smiling face, we know we are nearly done. Kimiya and Oleg do a close inspection of our gloves and take many pictures of them to send to the ground. The gloves are the most vulnerable parts of our suits, prone to cuts and abrasions, and the glove experts on the ground want to know as much as possible about how our gloves have fared today. Any holes will be easier to see while our suits are still pressurized.

When we are ready to get out of the suits, Kimiya helps us remove our helmets first, which is a relief in one way. But we will miss the cleaner air: the CO_2 scrubbers in the suits do a much better job than Seedra. Getting out of the suits was hard on Earth, but there we had the advantage of gravity, which helped by pulling our bodies down toward the floor. Here in space, my suit and I are floating together, so I need Kimiya to hold on to the arms of the suit and pull hard while he pushes down on the pants in the other direction with his legs. Extruding from the hard upper torso reminds me of a birthing horse.

Once I'm out of the spacesuit, it hits me all at once how draining it's been just being in the suit, never mind the full day of grueling work I did while wearing it. Kjell and I head to the PMM, where we remove our long underwear and dispose of our used diapers and biomedical sensors. We take a quick "shower" (move the dried sweat around on our bodies with wipes, then towel off to dry) and eat some food for the first time in fourteen hours. I call Amiko and tell her how it went—she watched the whole thing from mission control, but I know she's waiting to hear what it was actually like for me. She worried about this spacewalk more than she worried about any other part of this mission.

"Hey," I say as soon as she picks up the phone, "that was something. I don't know exactly how to describe it. It was fucking crazy."

"I'm so proud of you," she says. "It was intense to watch."

"It was intense for *you*?" I joke, though I understand what she means. She'd been in mission control since three in the morning Houston time and didn't eat or even go to the bathroom until I was back inside safely.

"It was more intense than seeing you launch," she says. "At least when you launched I had the chance to say good-bye to you right before. Today, I knew if something went wrong I would have to deal with not having seen you for seven months."

She tells me she was so excited for me that I was able to do a spacewalk after all these years of being an astronaut, and she says that everyone at NASA felt that enthusiasm.

"I'm beat," I say. "I'm not sure I want to do that again." I tell her that this was definitely the "type two" kind of fun—fun when it's done—but I know that by the time of our next spacewalk, I will be ready to go again. I tell her I love her before hanging up.

That evening, we go down to the Russian segment for a little celebration. Successful spacewalks are one of the events, like holidays, birthdays, and crew arrivals and departures, that warrant special dinners. This will be a short one, though, because Kjell and I are tired. While we eat, we talk about the day, what went well, what surprised us, what we might do differently next time. I tell Kjell what a great job he did, knowing he is still trying to put that errant switch throw behind him. He knows I don't give unearned praise, so I hope he can finish this day feeling that he did well. I tell Kimiya again what a great job he did as IV, and I thank the Russians again for their help. On days like this it's clear that this crew can truly pull together as a team, and that is one of the rewards of the hardest day I've ever had.

After we say our good nights, I slide into my bag, turn off the light, and try to fall asleep. As of tomorrow, Kjell, Kimiya, and Oleg will have spent one hundred days in space. Kjell and I will have some time to recover before preparing for our second spacewalk. That one will be even more complicated and physically demanding. But for now, I can rest. One of the biggest hurdles of this year is now behind me.

I CALL my father one evening to see how he's doing, and he tells me that my uncle Dan, my mother's brother, has died. He had suffered from a debilitating skeletal condition for most of his life, so his death was not a huge surprise, but because he was only ten years older than me he still seems too young to be gone. When Mark and I were about ten, Uncle Dan had moved into my family's basement for a while, and because he was closer to our age than to my mother's, I remember him being more like a big brother than an uncle. I remark to my father that death doesn't wait while I'm in space, any more than life does. The fact that I never said good-bye and won't be back until long after the funeral is a reminder that I'm missing things that can never be made up.

A few days later, I stop Kjell when he is floating through the U.S. lab and ask him if he could spare a minute. I put on a serious face and tell him I need to talk with him.

"Sure, what's up?" Kjell responds with his characteristic upbeat tone. People who are this sunny and positive can come across as fake, but I've learned from working with Kjell all this time in close quarters and under challenging circumstances that his attitude is completely real. He actually is that positive. I imagine this trait served him well as an emergency room doctor, and it's equally valuable in long-duration spaceflight.

"It's about the next spacewalk," I say, with a serious tone. I pause as if I'm searching for the right words.

"Yeah?" Kjell says, now with a hint of apprehension.

"I'm afraid I have to tell you—you're not going to be EV Two." EV2 was the role Kjell had played on our first spacewalk—I was the leader (EV1) as the more experienced astronaut, though it was the first time outside for both of us.

A look of concern crosses Kjell's face, followed quickly by sincere disappointment.

"Okay," he answers, waiting to hear more.

I decide I've fucked with him enough. "Kjell, you're going to be EV One."

It was a mean trick, but it's worth it to see the relief and excitement on his face when he realizes that he has been promoted. Kjell will fly more future missions and likely will conduct more future spacewalks, so it will be invaluable for him to get experience as the leader. I have full confidence in his ability to carry out this role, and I tell him so. We have a lot of preparation to do.

NOVEMBER 3 IS a midterm election day on Earth, so I call the voting commission in my home county—Harris County, Texas—and get a password that I can use to open a PDF they emailed to me earlier; I fill out my ballot and email it back to them. There are no political candidates on the ballot, just referendums. Still, I take pride in exercising my constitutional rights from space, and I hope it sends a message that voting is important (and that inconvenience is never a good excuse for failing to vote).

I follow the news from space, especially political news, and it seems like the presidential election next year is going to be like no other. Like the hurricanes I watch from above, a storm seems to be gathering on the horizon that will shape our political landscape for years to come. I pay close attention to the primaries of both parties, and though I don't tend to be a worrier, I start to worry. Sometimes before going to sleep I look out the windows of the Cupola at the planet below. *What the hell is going on down there?* I mutter to myself. But I have to concentrate on the things I can control, and those are up here.

16

THE RUSSIANS HAVE a very different system for medically certifying people to fly, and when we travel in their Soyuz we must abide by their rules. So it was a problem when my new flight surgeon Steve Gilmore presented me as a crew member to fly on the Soyuz to the International Space Station after having recently been treated for cancer.

Russian surgical procedures and treatment options for prostate cancer are not as advanced as those in the United States, and as a result, their statistics on survival and recovery are very different. Russian doctors overestimated the chances that I would experience debilitating negative effects from the surgery or have an early recurrence of the cancer. They were especially concerned that I would suddenly find myself unable to urinate in flight, which would require a costly and dramatic early departure. They didn't want to take that risk.

Steve worked hard to convince the Russian doctors that my surgery had been a success and that I was going to be able to pee just fine in space. We called Steve "Doogie," because of his youthful appearance, or "Happy," for his cheerful disposition. He worked on this issue for more than a year. It would have been easier for NASA to simply replace me with someone else, and I'm grateful they stood by me. In the end, the Russians agreed to let me fly, recognizing that our expertise and experience in this area were superior to theirs. They still made me fly with a catheterization kit in the Soyuz.

I began training for my mission to the space station in late 2007, with the launch scheduled for October 2010. Missions to ISS were divided into expeditions of six crew members, and my time on station would cover both Expeditions 25 and 26. In 2008, I began working with Sasha Kaleri, the Soyuz commander, and Oleg Skripochka, who would fly in the left seat as the flight engineer. Sasha is a quiet and serious guy with a full head of dark hair speckled gray. He was one of the most experienced cosmonauts, having flown three long-duration missions on *Mir* and one on the ISS—608 days altogether. He also brought a lot of old-school attitude and tradition, including some small Soviet flags, as part of his kit of personal items we get to launch in the Soyuz. He seems to be nostalgic for the Communist system, which of course was odd to me, but I liked him nonetheless. Oleg was on his first spaceflight. Studious and well prepared, he tried to model himself after Sasha in every way, and in turn Sasha treated Oleg like a son or a little brother.

This wasn't my first time training with the Russians, of course; I had trained to fly as the backup for Expedition 5 in 2001 and again as part of the backup crew for the flight prior to this one. By now, I was intimately familiar with the way the Russian space agency handles training similarly to NASA, such as an emphasis on simulator training, and the way they don't, like their emphasis on the theoretical versus the practical—to an extreme. If NASA were to train an astronaut how to mail a package, they would take a box, put an object in the box, show you the route to the post office, and send you on your way with postage. The Russians would start in the forest with a discussion on the species of tree used to create the pulp that will make up the box, then go into excruciating detail on the history of box making. Eventually you would get to the relevant information about how the package is actually mailed, if you didn't fall asleep first. It seems to me this is part of their system of culpability—everyone involved in training needs to certify that the crew was taught *everything* they could possibly need to know. If anything should go wrong, it must then be the crew's fault.

Before we can fly on the Soyuz, we must past oral exams, graded on a scale of one through five, just like the exams given throughout the entire Russian educational system. We took our final exams before a large commission, nearly twenty people in all, who were grading us.

We also had a larger audience of spectators. Privately, I referred to the oral exams as a "public stoning." Part of the process is a postexam debrief in which crew members argue for the grade they believe they have earned, minimizing and avoiding responsibility for any mistakes. This arguing over grades is something like a sport, and it seems we were being graded partly on how we pled our cases. I never wanted to argue—I was willing to take whatever grade the instructors wanted to give me, because I knew that in the end I would soon fly in space regardless.

Some of our training took place at other sites, as we learned to do everything from repairing the equipment on the space station to conducting experiments across many scientific disciplines. One day at the Johnson Space Center, I was in a session with a materials scientist who was teaching a group of astronauts how to use a new piece of equipment on the space station, a furnace for heating materials in zero gravity. While he was explaining the properties of the furnace, he showed us a golf-ball-sized sample of a material that had been "forged" in the furnace and said repeatedly that it was "harder than a diamond." I found that difficult to believe and asked whether I could hold it. He smiled and handed it to me.

"Is this really harder than a diamond?" I asked.

He assured me that it was.

I put the sample on the floor and raised my heel over it, looking at the scientist questioningly.

"Go ahead," he said.

I brought down my heel hard and the sample shattered, pieces flying all over the room. Apparently, it wasn't harder than a diamond. This incident became part of a narrative about me in some people's minds at NASA—that I didn't have enough respect for the scientific work being done on the space station. It's true that I'm not a scientist and that research was never my main motivation for going to space. But even if the science wasn't what drove me to become an astronaut, I have a profound respect for the pursuit of scientific knowledge and I take my part in that seriously. After all, I would argue, testing that sample from the furnace was an example of using the scientific method to gain knowledge.

Another uniquely Russian spaceflight practice was the creation of custom-molded seat liners for each crew member. The first time I served as a backup crew member, I went to Zvezda, the company that makes the Soyuz seats and Sokol suits, as well as the spacesuits the cosmonauts wear on spacewalks and ejection seats for Russian military aircraft. With a NASA flight surgeon and an interpreter who specializes in medical translation, I traveled to the other side of Moscow from Star City, through many miles of Moscow suburbs. Once inside the guarded and gated Zvezda facility, I was helped into a container like a small bathtub, then had warm plaster poured in all around me. After the plaster hardened, I was helped out, then got to watch while an old weathered technician with a beard like Tolstoy's—he was really more like an artisan—went to work. I watched his huge, callused hands, with long sensitive fingers like a sculptor's, carve out the excess plaster to create a perfect mold of my back and butt.

A few weeks later, I came back to Zvezda for a fit check of the newly made seat liner, followed by the dreaded pressure check—an hour and

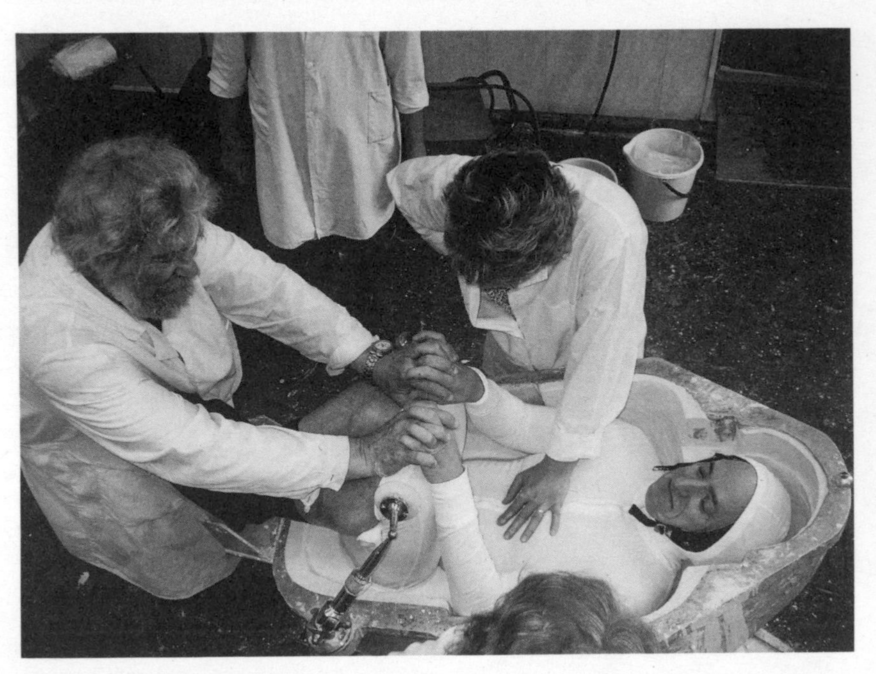

Pouring plaster for my Soyuz seat liner

a half on my back in my custom-made spacesuit in my custom-made seat liner with the suit pressurized. The circulation in my lower legs got cut off, and the position became a distinctly painful form of torture. All the cosmonauts and astronauts dread this procedure, but if anyone complains, they are met with a curt answer: "If you can't deal with this pain now, how can you deal with it in space?" I never bothered arguing about it, but this was a flawed argument; in space you can deal with discomfort that you know is keeping you alive. A few weeks later, I came back for the pressure check again, this time in a vacuum chamber, a ritual meant to give us confidence in the suits. These activities can feel more like ceremonial rites of passage than engineering necessities, as with so many of the traditions in the Russian space program. In the coming years, I would carry out this painful ritual two more times.

We traveled to Baikonur two weeks before the scheduled launch date. The last morning, we went through the process of getting suited up, doing our leak checks, and speaking to our loved ones through the pane of glass. We rode the bus out to Gagarin's launchpad, peed on the tire, and climbed into the capsule. Among the things we had to do to get the vehicle ready was configure the oxygen system, a task that was the responsibility of the flight engineer 2—in this case, me. Toward the end of the launch countdown, I was manipulating one of the O_2 valves when we heard a loud squeal. We guessed it was compressed oxygen leaking into the cabin, and we were right. I immediately closed the valve, but we had a massive oxygen leak anytime the valve was taken out of the closed position, which was a requirement during the flight.

At the direction of the ground, Sasha tried to get the situation under control by venting the O_2 through the hatch valve into the habitation module above us, then overboard through a valve that led to the outside. He unstrapped himself from his seat so he could sit up to reach the valve directly over his head. I looked at the readings on our LCD screens, paying close attention to the partial pressure of oxygen compared to our total pressure. I did some mental math and calculated that there was close to 40 percent oxygen in the capsule, the oxygen concentration threshold at which many materials become easily flammable with an ignition source, like a small spark.

All astronauts knew that the crew of Apollo 1 had been killed by a fire in their capsule because it had been pumped full of pure oxygen and that a tiny spark caused a conflagration of the Velcro-lined capsule. NASA didn't use high-pressure oxygen anymore in this way, and they also redesigned the hatch on the Apollo capsule to open outward—and so have all hatches on NASA human-rated launch vehicles since. Not the Russians. The hatch on our Soyuz opened inward, so if there was a fire, the expanding hot gases would put outward pressure on the hatch and trap us, just like the crew of Apollo 1. As Sasha struggled to reach the valves, flailing in his seat, the metal buckles on his straps struck exposed metal on the capsule's interior. I thought distinctly to myself, *This is not a good place to be right now.*

Once Sasha was back in his seat and it seemed clear we weren't going to catch fire, we talked about our predicament. I decided not to voice my concern about the flammability risk.

"It's too bad we won't launch today," I said.

"*Da*," Sasha agreed. "We will be first crew to scrub after strapping in since 1969." This is an incredible statistic, considering how often the space shuttle used to scrub, right up to the seconds before launch, even after the main engines had lit.

A voice from the control center interrupted us. "Guys, start your Sokol suit leak checks."

What? Sasha and I looked at each other with identical What-the-fuck? expressions. We were now inside five minutes to launch. Sasha raced to get strapped back into his seat properly. The emergency escape system had been activated, and if something had set it off, the rocket would have launched us away from the pad without warning. Sasha would have likely been killed if it had activated while he was not strapped in. We closed our visors and rushed through the leak check procedures. With less than two minutes to spare, we were ready to go. We settled into our seats for our last minutes on Earth.

The launch experience was different from the shuttle—the Soyuz capsule was much smaller than the shuttle's cockpit, and less advanced, so there was less for the crew to do. Still, it was much more automated than the space shuttle was. Nothing could match the acceleration of the shuttle's solid rocket boosters pushing us away from the Earth with an instantaneous 7 million pounds of thrust at liftoff, but anytime you rocket off the planet, it's serious business.

Once we reached orbit, we were stuck in this cold tin can with very little to do for two whole days until we were to dock to the space station. As the spacecraft moved in and out of communication coverage, the sun rising and setting every ninety minutes, we quickly lost track of any normal sense of time and drifted in and out of sleep. The habitation module was cramped and spartan, lined in a dull yellow Velcro with an occasional exposed metal frame or structure, which quickly became covered in condensation. We didn't even have a good view of the Earth because the Soyuz constantly spun to keep the solar arrays pointing at the sun in order to charge its batteries. I had brought my iPod, but the battery soon died. I spent most of the time floating in the middle of the habitation module, feeling like I did when I was a kid in after-school detention, staring at the clock, waiting for the day to be over. When docking day came, I was excited, but when I looked at my watch and realized the moment we were to float through the hatch was still eighteen hours away, I thought to myself, *Oh, shit. What the fuck am I going to do for the next eighteen hours?* The answer is: nothing. I just floated there. I've said that any day in space is a good day, and I believe it, but two days in a Soyuz is not *that* good.

This was also Amiko's first time seeing me launch into space. She had witnessed three previous shuttle launches, including one of my brother's, before I knew her well. (She remembered seeing me at a prelaunch party in Cocoa Beach, Florida, carrying around a sleeping baby Charlotte with her head of curly blond hair.) So she wasn't new to the launch experience, but it was different traveling to Baikonur and seeing the Russian way of doing things. And of course it was very different seeing a launch with someone she cared about on board. My brother told me later, once I was safely in orbit, that she cried while watching my launch. I was surprised to hear that because, despite the fact that we had been together for more than a year, I had never seen her cry. When I asked Amiko about it, she said she hadn't been expecting to be so emotional, but she was moved by the beauty and awe of the launch and by her happiness for me. She knew what it meant to me to get to fly in space, and she knew how hard I had worked to get there.

Years later, I learned more about what went on that day in the launch center in Baikonur. Someone in launch control had said that they understood this anomaly, and that there was a workaround: cycling the oxygen valve partially open, then closing it before opening it fully to reseat a sticky valve. In the minutes before launch, officials were passing around a piece of paper they needed to sign indicating that they were go for launch in spite of the oxygen leak and Sasha's struggle to equalize the pressure as time ticked away. As a crew member getting ready to ride the rocket to space, I found this to be troubling.

WHEN I FLOATED through the hatch to officially join the crew of Expedition 25 on the ISS, I was elated to be starting a long-duration mission. It had been a long road from DOR, backup on Expedition 5, the *Columbia* accident, my STS-118 mission, prostate cancer, my second backup training flow, and now my prime assignment—ten years all told.

On board were two Americans and one Russian: Doug Wheelock was serving as the commander for this expedition and would be turning over ISS to me when he left. Doug was a great first ISS commander to serve under. He took a hands-off approach to leadership, letting everyone find his or her own strengths.

My other American crewmate was Shannon Walker. I didn't know Shannon very well before this mission, but I was surprised by how different she looked when we met in space: her hair had grown out gray over the months she'd been in space without access to hair dye. Shannon had trained to fly in the left seat of the Soyuz, which meant she needed to know the systems well enough to take over in an emergency that might incapacitate the Russian Soyuz commander, and as a result she spent much more time training in Russia than I did. When I got on board ISS, I was impressed with her abilities as a crew member. This was her first flight, so when I arrived I was at first thinking of her as a rookie, but it didn't take long for me to realize that she had almost ten times as much time in space as I did, and that in fact I could use her help. At NASA, we talk about "expeditionary behavior," which is a loose term for being able to take care of yourself, take care of others, help out when it's needed, stay out of the way when necessary—a combination of soft skills that's difficult to define, hard to teach, and a significant challenge when they are lacking. Shannon was a master at this.

The Russian cosmonaut Fyodor Yurchikhin, a short, stocky guy with a broad smile, was already on board. Fyodor was one of only two people I've been in space with more than once (the other being Al Drew). Fyodor was born in the country of Georgia to Greek parents, which is unusual in a cosmonaut corps made up largely of ethnic Russians. He had a real enthusiasm for photography, and loved taking pictures of the Earth. Even more, he loved showing his pictures to his crewmates regardless of what they were trying to do at the time. The cosmonauts aboard ISS generally don't have as hectic a schedule as the Americans do, and sometimes that difference can show when they are free to socialize during the day, floating around the dining room table sharing coffee or a snack, while we rush from one thing to the next.

On this mission I learned the differences between visiting space and living there. On a long-duration flight, you work at a different pace, you get more comfortable moving around, sleep better, digest better. As my first long-duration mission went on, what surprised me most was how little force it actually took to move around and to hold myself still. With just a slight push of a finger or a toe, I could travel across a module and wind up exactly where I wanted to be.

One of the first tasks I tackled when I got on board was repairing a device called Sabatier, which combines oxygen from the CO_2 the Seedra collects and leftover hydrogen from the oxygen-generating assembly to create water. Sabatier is an important part of the nearly closed-loop environmental system of the station. My job was to tune the system, a tedious multiday task, using flow meters and other diagnostic tools. At the time I thought I handled it well, but looking back at it years later with much more experience doing this kind of repair, I realize how much Shannon helped me by getting all the tools and parts together for me in advance, checking on me when I seemed to be struggling, and encouraging me when I got frustrated. Without her help, the task would have been nearly impossible to carry out so early in the mission.

I celebrated my first Thanksgiving on the space station shortly before taking over as commander of ISS for the first time. The next day, Shannon, Doug, and Fyodor departed for Earth, leaving behind Sasha, Oleg, and me.

A few weeks later, the new crew arrived. American astronaut Cady Coleman was a retired colonel in the U.S. Air Force and held a Ph.D. in chemistry, and, I came to learn, played the flute. Some people who knew Cady and me thought that we might not click as crewmates, or that I might kill her, because we came from such different backgrounds—the fighter pilot (me) and the scientist (Cady). In fact, Cady and I became great friends and she was a great crew member, even though I was never able to get her to go to bed on time. Sometimes I would get up to use the bathroom at three a.m. on a work night and find her playing her flute in the Cupola. Cady taught me how to be more in touch with my feelings and those of the people we worked with on the ground. She also helped me see the value of reaching out more to the public, letting people share the excitement of what we were doing in space. This would turn out to be enormously helpful on my yearlong mission.

The Italian astronaut Paolo Nespoli, a talented engineer with a great sense of humor, was the third member of the new crew. Paolo is really tall—too tall to fit into the Soyuz, in fact, and the European Space Agency had to pay the Russians to modify the seat, setting it at a steeper angle, in order to fit him into the capsule.

The Soyuz commander was Dima, who had been on the Expedition 5 backup crew with me and with whom I had done survival training a decade ago. This was his first spaceflight. Back when Dima and I were assigned to the same backup crew, he had argued that he should be commander of ISS because he was the commander of the Soyuz and a military officer. Sasha Kaleri, with his extensive experience in space, was much better qualified to command but was not a military officer. Dima was so convinced that he had been wronged that he wrote two strongly worded letters to his management saying that Sasha wasn't performing up to par and should be removed from the crew. This incredible breach of protocol meant that Dima was not assigned to a flight for many years, despite his superior technical skills.

I had heard about crew members not getting along during spaceflights, but I had never personally experienced it myself—until now. I floated down to the Russian segment one day to ask about something, and while I was there Dima asked for my help with a piece of Russian hardware he was struggling to fix—the Russian Elektron, their device for producing oxygen from water. That wouldn't have seemed unusual, but he asked me with Sasha floating in the vicinity. Sasha offered to help, to his credit, but Dima pretended not to hear him. I couldn't imagine what it was like working, eating, and sleeping on top of each other for four months with that much tension between them. Their lack of communication made their work harder and could have cost them their lives—and, potentially, ours too—in an emergency.

After I had been in space for a few months, the press was reporting that Sasha Kaleri had brought with him a Quran that had been given to him by Iran. The rumor was that the Quran was a symbolic response to a recent desecration of Qurans in the United States on the anniversary of the 9/11 terrorist attacks. The ISS program manager wanted to know if it was true. When the chief astronaut asked me about it, I said I didn't care what books crew members brought on board with them, and I was surprised that NASA would take an interest in such details. I said I wasn't going to ask anyone about his or her private belongings, and I thought that would be the end of it. But soon after, I heard directly from the space station program manager: I was told in no uncertain terms that I was expected to find out whether Sasha had brought a Quran on board.

Usually I would push back against a request from the ground only once, and if they persisted, I would do it their way unless it was a safety issue. This was easier than having a showdown over every small disagreement and would preserve my sanity and energy for when it was really needed. But in this case, I still felt strongly that I shouldn't acquiesce.

The next day, I floated over to the Russian segment and found Sasha in the confined space of the Russian airlock, working on one of their spacesuits.

"Hey, Sasha," I said. "I'm supposed to ask you something, but I don't personally care what the answer is."

"Okay," Sasha said.

"I'm supposed to ask you whether there's an Iranian Quran on board the station."

Sasha thought for a moment. "That's none of your business," he said agreeably.

"Got it," I responded. "Take it easy." I floated back to the U.S. segment and passed his answer on to my management. That was the last I heard of it.

JANUARY 8, 2011, WAS a bright sunny day in Tucson, Arizona, but on the space station, the weather was the same as always, and I was fixing the toilet. I had taken it apart and organized the pieces around me so they wouldn't float away, and now I would not do anything else until I finished the job. We can use the toilet in the Russian segment if necessary, but it's far away, especially in the middle of the night, and puts unnecessary stress on their resources. The toilet is one of the pieces of equipment that gets a great deal of our attention—if both toilets break we could use the Soyuz toilet, but it wouldn't last long. Then we would have to abandon ship. If we were on our way to Mars and the toilet broke and we couldn't fix it, we would be dead.

I was so involved in the work that I didn't notice the TV feed being cut. We lost our signal pretty routinely, whenever the space station went out of the line of sight between our antennas and the communication satellites, so I didn't think it was a big deal. Then a call came from the ground.

Mission control told me that the chief of the Astronaut Office, Peggy Whitson, needed to talk to me and would be calling on a private line in five minutes. I had no idea why, but I knew the reason wouldn't be anything good.

Five minutes is a long time to think about what emergency might have occurred on the ground. Maybe my grandmother had died. Maybe one of my daughters had been hurt. I didn't make any connection between the blank TV screen and the phone call—NASA had deliberately cut the feed to spare me learning bad news.

Before leaving for this mission, I had decided that Mark should act as my proxy in cases of emergency. He knew me better than anyone, and I trusted him to decide what I should hear and when, whether it should come through him or someone else, like a flight surgeon or another astronaut. He knew that in a crisis I would likely want to have all the information up front as soon as possible.

Peggy came on the line. "I don't know how to tell you this," she said, "so I'm just going to tell you. Your sister-in-law, Gabby, was shot."

I was stunned. This was such a shocking thing to hear, it seemed surreal. Peggy said she didn't have any more information, and I told her I wanted to know any news as it came in, that she shouldn't keep secrets to spare me. Even if the information was unconfirmed or incomplete, I still wanted to know.

When I got off the line, I told Cady and Paolo what had happened, and then I told the cosmonauts. I tried to assure everyone I was going to be okay, but I also told them I was going to need some time and that I was going to spend most of it on the phone. They were shocked and upset as well and of course gave me the room I needed. Though I was hesitant to turn over this crucial job of fixing the toilet to Cady and Paolo, I had no choice but to trust them.

I liked Gabby from the first time I met her, and I've only gotten to like her more over the years. She treats everyone the same—she is interested in everyone she meets, no matter who they are, where they are from, or what political party they vote for. She wants to help everyone she comes across, and she was completely dedicated to her work as a congresswoman on behalf of the people of Arizona. That was why it was so hard to fathom what had occurred. This sort of random violence should never happen to anyone, but it seemed especially awful that this should happen to her.

I called Mark. He was hurriedly packing his bags in Houston as we talked and had arranged to fly to Tucson as quickly as possible. He told me he'd received a call from Pia Carusone, Gabby's chief of staff, telling him about the shooting. Pia told him that Gabby had been shot at a public event, that an unknown number of people had been hurt or killed, that Gabby's status was uncertain, and that he needed to get to Tucson right away. Mark said okay and hung up—then immediately called Pia back and asked her to repeat everything she had just said. The idea of his wife being shot was so shocking, it simply hadn't sunk in. He needed to hear Pia say it all over again to be sure that it was real.

Mark and I agreed we would connect again as soon as he landed in Tucson. Not long after, mission control called to tell me that the Associated Press was reporting that Gabby had died.

I immediately tried to call Mark again, but he was already in the air on the way to Tucson with our mother and his two daughters. Our good friend Tilman Fertitta had lent them his private jet so they could get to Tucson as quickly as possible. This is the kind of thing Tilman does for his friends, and I've always been grateful for the way he stepped up that day. I called Tilman to find out what he had heard.

"Gabby's not dead," he said. "I don't believe it."

"How do you know that?" I asked. "All the news media are saying it."

"I don't know for sure, but it just doesn't make any sense," he said. "She went into surgery, and she should still be in the operating room."

One of the things I like about Tilman is that he immediately sees through bullshit. Even with a subject completely outside his expertise, like brain surgery, he questions everything, and most of the time he's right. I took hope in his words.

The next several hours were some of the longest in my life. My mind kept traveling to my brother—what he must be feeling, not knowing whether he would ever see his wife alive again. I called Amiko and my daughters and repeated to them what Tilman had told me: regardless of what they were saying on TV, it didn't make sense that Gabby was dead. Soon after Mark landed in Arizona, I got him on the phone.

"What's happening?" I asked the moment he picked up. "They're saying Gabby died."

"I know," he said. "I got the news on the plane. But I just spoke to the hospital, and it was a mistake. She's still alive."

There is no way to describe the relief you feel when you've been told someone you care about is alive after spending hours thinking she was dead. We knew Gabby would still have a long and hard road ahead of her, but knowing she was still drawing breath was the best news we could have hoped for.

I made dozens more calls that day and the next—to my brother, to Amiko, to my mother and father, to my daughters, to friends. Sometimes I wondered whether I was calling too much, whether in my effort to be there for them I was becoming intrusive. That first day, I learned that thirteen other people had been injured in the shooting and six had been killed, including a nine-year-old girl named Christina-Taylor Green, who was interested in politics and wanted to meet Gabby. I talked to Mark and Amiko a dozen times that day.

The next day, we had a videoconference with Vladimir Putin that had been planned long before. I was surprised by how much of the time he took speaking directly to me. He told me that the Russian people were behind my family and that he would do anything he could to help. He seemed sincere, which I appreciated.

On Monday, President Obama announced a national day of mourning. The same day, I was to lead a moment of silence from space. I don't get nervous easily, but this responsibility weighed heavily on me. This would be the first public statement from my family. As the time grew near, I called Amiko at work in mission control in Houston. I confided to her that I was uneasy—I didn't know exactly how long a moment of silence should be, and for some reason I had focused on that seemingly insignificant question.

"It should be as long as you want," she assured me. "As long as feels right."

Her assurance helped. At the appointed time, I floated in front of the camera. I'd put a great deal of thought into the brief remarks I'd written, but I wanted to sound as if I was speaking from the gut rather than reading from a prepared statement—because I was.

"I'd like to take some time this morning to recognize a moment of silence in honor of the victims of the Tucson shooting tragedy," I said. "First, I'd like to say a few words. We have a unique vantage point here aboard the International Space Station. As I look out the window, I see a very beautiful planet that seems inviting and peaceful. Unfortunately, it is not.

"These days, we are constantly reminded of the unspeakable acts of violence and damage we can inflict upon one another. Not just with our actions, but with our irresponsible words. We are better than this. We must do better. The crew of ISS Expedition Twenty-six and the flight control centers around the world would like to observe a moment of silence in honor of all the victims, which include my sister-in-law, Gabrielle Giffords, a caring and dedicated public servant. Please join me and the rest of the ISS Expedition Twenty-six crew in a moment of silence."

Those of us who have had the privilege to look down on the Earth from space get the chance to take a larger perspective on the planet and the people who share it. I feel more strongly than ever that we must do better.

I bowed my head and thought of Gabby and the other victims of the shooting. Just as Amiko had assured me, it wasn't hard to sense when the moment was complete. I thanked Houston, and we went back to work for the day. On the space station, we followed our normal routine. But I knew that on Earth some things would never be the same.

My brother had been assigned to the second-to-last flight of the space shuttle program, a mission to deliver components to the International Space Station. He was scheduled to fly on April 1, less than three months after the shooting. Gabby's condition was stable, but she had a long road of surgeries and therapy ahead of her. He knew that if he was going to step aside and let someone else command that mission, he should do it as soon as possible so the new commander would have time to get up to speed.

It wasn't clear whether NASA management would make the decision or if Mark would be allowed to decide for himself, and the uncertainty added to Mark's stress. And in the early days after the shooting, he wasn't sure what he would choose if it were up to him. He wanted to be there for Gabby as she started the long process of recovering from her catastrophic injury, but he also felt a duty to see his mission through. He and his crewmates had been training together for months, and a new commander would not know the mission or the crew as well as Mark did. We talked on the phone about it many times, but in the end, it was Gabby who made the decision. She would have been devastated if the shooting had robbed him of his last chance to fly in space. She urged him to go.

Because astronauts always have to prepare for the possibility we won't survive a mission, Mark now had to consider his responsibilities to Gabby in a new way. On his previous flight, Mark had straightened out his affairs and written a letter to be delivered to Gabby in the eventuality he didn't come back. But now he wasn't just Gabby's husband, he was also her main caretaker and her main support. If he were suddenly gone, that would be disastrous in a very different way.

In all our discussions about Mark's mission and the possibility he could die, we couldn't help but note the irony. As astronauts, Mark and I had been confronted with the risks of flying in space. None of us could have imagined that it would be Gabby, not Mark, whose work nearly cost her her life.

IN FEBRUARY, *Discovery* came up and docked with the station on its last flight. It was funny seeing a shuttle crew come on board, as I had done myself not that long before, flying around in the Superman orientation—floating horizontally—rather than the more upright position of seasoned long-term space travelers. The new guys were bumping into everything and kicking equipment off the walls everywhere they went. The day after they arrived, I floated over to visit *Discovery,* the orbiter I had flown on my first flight. I remembered the last time I had been on board, December 28, 1999, climbing out that evening at the end of the flight. It seemed like a lifetime ago.

Looking around the mid-deck, I felt a wave of nostalgia for the time I'd spent there. The three remaining orbiters were remarkably similar, especially *Discovery* and *Endeavour,* which both had similar upgrades made after *Columbia* had been assembled. But there were differences between them that we who flew on them found unmistakable. When I had flown on *Endeavour* several years earlier, it still had a new-car look to it, even though it had been in service for sixteen years. At twenty-seven, *Discovery* was the oldest orbiter, the workhorse of the fleet on its thirty-ninth and last mission. But rather than seeming past its prime, it seemed to me like a well-loved classic with the refinement of a vintage car.

I floated up to the flight deck where the pilot, Eric Boe, was strapped into his seat running through some checklists. He greeted me and continued with his work.

"Hey, Eric, do you mind if I jump up into your seat for a minute? I just want to see how it feels."

Eric is a sharp guy. "Your first flight was on *Discovery,* wasn't it?" he asked. With a smile, he got out of the way.

I floated into the seat and strapped myself in, surveying my former office. I looked over the mass of switches, buttons, and circuit breakers that controlled the many complex systems I had been responsible for so many years ago. I knew I could fly her right now if I had the chance. I remembered sitting in this seat, and my former self seemed so young and inexperienced compared to where I was now, having spent so much more time in space. Little did I know what the future still had in store for me.

The crew of the *Discovery* mission did a few spacewalks, and one of them involved a Japanese payload called "Message in a Bottle." It wasn't an experiment but simply a glass bottle that Al Drew opened at a certain point in the spacewalk to "let space in." Once the glass bottle returned to Earth, it would be displayed in museums throughout Japan in order to raise children's interest in spaceflight (personally, I was skeptical about how excited children would get about an empty glass bottle). After the spacewalk was over and the bottle was back in the station, the Japanese control center wanted to know whether I had "safed" the bottle (I was supposed to tape the lid closed to make sure it wasn't accidentally opened). I was busy doing a number of things, but they kept pestering me about it, until I finally got on the Space to Ground channel and said, "Message in a Bottle is on *Discovery,* and I opened it to make sure there was nothing inside." There was a long pause. Then I said, "Just kidding."

Shortly after *Discovery* and its crew returned to Earth, it would have its engines removed and would be sent to the Smithsonian's National Air and Space Museum in Washington, D.C., to become a permanent display. *Discovery* had left Earth more times than any spacecraft in history, a record of distinction that I expect it to keep for a long time to come.

Sasha, Oleg, and I were scheduled to return to Earth on March 16, 2011. Never having experienced reentry on the Soyuz, I was curious what it would be like. For some strange reason that I still haven't quite figured out, people didn't talk about the experience of landing in the Soyuz as much as we did with the space shuttle. It might partly be because, until recently, the astronauts who flew on Soyuz were not former test pilots and so maybe lacked the curiosity bordering on obsessive interest in the performance of spacecraft that shuttle pilots had. I had been given a range of different impressions by different people—that it would be terrifying, that it would be fine, and that it would actually be fun, like Mr. Toad's Wild Ride at Disney World.

That day everyone was concerned about the weather, because there were blizzard conditions at the landing site. Our capsule smacked into the hard frozen surface of the desert steppes of Kazakhstan, bounced around, tipped over, then was dragged a hundred yards by the parachute. I've never been in a car accident that ended in multiple rollovers, but I imagine that landing in the Soyuz that day felt a lot like that—violently jarring. But I found it exhilarating.

Eventually, the rescue forces corralled the parachute and knocked it down before it could drag us any farther. Not long after, the hatch opened and the blizzard blew into the capsule—the first fresh air I had smelled in six months, incredibly refreshing. It was a sensation I'll never forget.

. . .

A FEW DAYS AFTER I returned to Earth, Amiko and I went to visit Gabby at TIRR Memorial Hermann, where she was being treated. I was shocked at first by how different she looked. She was in a wheelchair and was wearing a helmet to protect her head where a piece of her skull had been removed to allow her brain to swell. Her hair was short—it had been shaved for brain surgery—and her face looked different. It took me a moment to process the enormity of what had happened to her. When I heard she had been shot, I had understood intellectually what that meant. But it was another thing entirely to see my vivacious sister-in-law in such a different state—not only physically changed but unable to speak as she once had. Sometimes Gabby would get a look on her face as if she had something to say, and when we all looked at her and paused, she would say, simply, "Chicken." Then she'd roll her eyes at herself—that wasn't what she wanted to say!—and try again.

"Chicken."

I could see how frustrating it was for Gabby, who used to make speeches to thousands of people that inspired them and won their votes. Mark explained that she had aphasia, a communication disorder that made it hard for her to speak, though her ability to comprehend language, her intelligence, and, most important, her personality remained unaffected. She understood everything we said to her, but putting her own thoughts into words was extremely challenging.

We had dinner together at the hospital, and as the visit went on I could see Gabby's warmth and her sense of humor. Later, Amiko and I talked about how Gabby was doing, and Amiko said she thought Gabby looked great considering how recent her injury was. Amiko reminded me how long it had taken her own sister to walk, to speak, to regain aspects of her personality after sustaining a traumatic brain injury in a car accident. Amiko didn't want to be overoptimistic, but she knew from experience that people could make huge strides after being in terrible shape. Gabby was still very much herself, and that was a promising indication for her recovery.

"I can see Gabby in Gabby" is how Amiko put it, and she was right.

Less than two months later, I was standing next to Gabby on the roof of the Launch Control Center at Kennedy, watching *Endeavour* prepare to launch for its last mission, with Mark as its commander. Gabby had been to a space shuttle launch before, and of course I had been to many. It's an experience that never gets old. The ground shakes, the air crackles with the power of the engines, and the rockets' flames burn a searing orange in the sky. Seeing an object the size of a tall building lift itself straight up into the sky at supersonic speed is always moving, and when someone you know and care about is on board, it's that much more so. There was a low cloud cover that day, and *Endeavour* punched up through it, lighting them up orange for a moment, then disappeared. Eight minutes later, it was in orbit around the Earth.

When Mark had decided he would see this mission through, Gabby had set the goal of being well enough to fly to Florida and see him off. That had been extremely ambitious, and she had done it. For Gabby, just being here was an accomplishment on par with a shuttle launch. She seemed to thrive on the challenge to do hard things.

Soon after, the space shuttle was retired, fulfilling the terms of the *Columbia* Accident Investigation Board. I was sad to see it go. The shuttle was unique in its range of capabilities—heavy-lift cargo vehicle, science laboratory, orbiting service station for busted satellites. It was the spacecraft I'd learned to fly and learned to love, and I doubt we'll see anything like it again in my lifetime.

In 2012, NASA learned the Russians were going to send a cosmonaut to the space station for a year. Their reasons were logistical rather than scientific, but once this had been decided, it put NASA in the position of having to either explain why an American astronaut was not up to the same challenge or announcing a yearlong mission themselves. To their credit, they chose the latter.

Once the Year in Space mission had been announced, NASA still had to choose the astronaut to do it. At first, I wasn't sure I wanted it to be me. I remembered exactly how long 159 days on the space station had felt. I had spent six months at sea on an aircraft carrier and that was long; six months in space is longer. Spending twice as long up there wouldn't just feel twice as long, I thought—it could be exponential. I knew I would miss Amiko and my daughters and my life on Earth. And I knew how it would feel if something bad happened to someone I loved while I was gone, because I had already experienced it. My father was getting up there in age and wasn't in the greatest health.

But I had decided a long time before always to say yes to whatever challenge came my way. This yearlong mission was the hardest thing I'd ever have the opportunity to do, and after some reflection I decided I wanted to be the one to do it.

Many other astronauts also expressed interest. After all, spaceflight opportunities don't come around every day. The requirements to be considered were many: we had to have previously flown a long-duration flight, we had to be certified to do spacewalks, we had to be capable of being assigned as commander, we had to be medically qualified, and we had to be available to be off the Earth for that year. With such a fine filter put on the requirements, in the end only two people qualified: Jeff Williams, one of my astronaut classmates, and me.

At around the same time, NASA was also looking to assign a new chief astronaut, because Peggy Whitson had stepped down in order to be eligible for the yearlong mission herself. I put my name in for the job of chief. In the interview, I was asked whether I would rather be chief of the Astronaut Office or fly in space for a year. Without hesitation, I said, "Chief of the Astronaut Office." I thought there would be other opportunities to fly in space again, but maybe not another opportunity to serve as chief astronaut. My preference might have been considered, but the managers decided differently. A few weeks later, I learned I would be flying in space for a year.

Twenty-four hours after I was assigned, I was told that after further evaluation I had been medically disqualified and that Jeff would fly. On my previous mission I had experienced some damage to my eyes, and NASA didn't want to take the risk of sending me up again. There could be some unexpected acceleration of the harmful effects after the six-month point, leaving me with permanent damage to my vision. I thought the danger was exaggerated, and I was disappointed, but I was resigned to the decision.

When I got home that night, I told Amiko about being medically disqualified. Rather than looking disappointed, as I expected, she looked puzzled.

"So they're going to send someone who has been on two long flights and has *not* suffered vision damage?" she asked.

"Right," I said.

"But if the point of this mission is to learn more about what happens to your body on a long mission," she asked, "why would they send someone who is known to be immune to one of the things they intend to study?"

This was a good point.

"In all the time I've known you," she said, "I have never seen you take no for an answer so easily."

That night, after Amiko was asleep, I looked through my NASA medical records, an enormous pile of paper two feet high documenting years of data. I had experienced damage to my vision on my long-duration flight, but it had been mild and had returned to normal when I came back to Earth, though I still had some structural changes. Amiko was right: we could learn more about vision changes from someone like me than from someone who had demonstrated an immunity to the problem. I decided to present my case to management. They listened, and to my surprise they reversed their decision.

When I was preparing for the press conference to announce Misha and me as the one-year crew members, I asked what I thought was an innocent question about genetic research. I mentioned something we hadn't previously discussed: Mark would be a perfect control to study throughout the year. It turns out my mentioning this had enormous ramifications. Because NASA was my employer, it would be illegal for them to ask me for my genetic information. But once I had suggested it, the possibilities of studying the genetic effects of spaceflight transformed the research. The Twins Study became an important aspect of the research being done on station. A lot of people have assumed that

I was chosen for this mission because I have an identical twin, but that was just serendipitous.

The yearlong mission was announced in November 2012, with Misha and me as the crew.

THE IDEA of leaving the Earth for a year didn't feel especially vivid until a couple of months before I was to go. On January 20, 2015, I attended the State of the Union Address at the invitation of President Obama. He was planning to mention my yearlong mission in his speech. It was an honor to sit in the House Chamber with the gathered members of Congress, the Joint Chiefs, the cabinet, and the Supreme Court. I sat in the gallery wearing my bright blue NASA flight jacket over a shirt and tie. The president described the goals of the yearlong mission—to solve the problems of getting to Mars—and called me out personally.

"Good luck, Captain!" he said. "Make sure to Instagram it! We're proud of you."

The assembled Congress got to their feet and applauded. I stood and gave an awkward nod and a wave. To see the government come together, even if only in a physical sense, was touching, and it was great to experience in person the bipartisan support NASA often enjoys.

I was seated next to Alan Gross, who had been held in a Cuban prison for five years. He suggested that while I was in space I should count up—count the number of days I had been there—rather than counting down the number of days I had left. It will be easier that way, he said. And that's exactly what I did.

17

November 6, 2015

Dreamed I was back on Earth and was allowed to return to the Navy to fly F-18s off the aircraft carrier. I was elated because I thought I would never get to fly like that again. I went back to my old squadron, the World Famous Pukin' Dogs, and all the same guys were there, unchanged from when I left. It was great because I was allowed to be like a junior officer even though I have the rank of captain. Because I had so much flight experience, everything was easy for me, supernaturally easy, especially landing on the ship.

NOVEMBER MARKS the nine-year anniversary of my surgery, and I reflect on the fact that I have spent more than a year of my life in space *after* having been diagnosed with and treated for cancer. I don't think of myself as a "cancer survivor"—more like a person whose prostate gland had cancer, which was removed and disposed of. But I'm happy if my story is meaningful to others, especially kids, as an example that they can still achieve great things.

Once again, Kjell and I have spent days preparing our suits and equipment, reviewing procedures, and conferencing with experts on the ground. This spacewalk will have two goals: one is to replumb a cooling system, bringing it back to its original configuration so a spare radiator can be saved for future use. The other is to top off that cooling system's ammonia supply (the space station uses high-concentration ammonia to cool the electronics). These tasks might sound unexciting, and in many ways they are. Yet the story of how we have kept the space station cool—a huge chunk of metal flying through space getting roasted by the unfiltered sun for forty-five minutes out of every ninety while its enormous solar arrays generate electricity—is a story of an engineering triumph with important implications for future spaceflight. The work Kjell and I will do today to keep the cooling system working will be one small piece of that larger story, just as the work of the astronauts and cosmonauts who have performed the hundreds of spacewalks from the station over the years have each contributed something invaluable to its construction.

The day of our second spacewalk starts much like our first: up early, quick breakfast, prebreathe oxygen, get suited up. Today I've decided to wear my glasses, because I found the Fresnel lenses attached to my visor didn't work as well as I'd hoped on our first venture outside the station. At one point, the tether for one of the tools I was working with became tangled, and I wasn't able to see the knot clearly enough to undo it. Luckily, it magically untangled itself. There are risks involved in wearing glasses—if they slip off, there will be nothing I can do about it with my helmet on, but I prepare for that problem by taping them to my head. Being bald, I have the perfect haircut for this technique. I regret not getting comfortable with contact lenses.

I put my comm cap on and scratch any itchy spots one last time before my helmet is sealed. Kjell and I get into the airlock. This time, I know that neither of us will flip our water switches early, and I also know I won't have to be the one to struggle to get the hatch open or closed—that's the job of the lead spacewalker.

Our work site today will be all the way out at the end of the truss, 150 feet from the airlock—so far that we need to use the length of both our safety tethers together to reach it. As we start the journey, translating hand over hand along the rails, I notice again how much damage has been done to the outside of the station by micrometeoroids and orbital debris. It's remarkable to see the pits in the metal handrails going all the way through like bullet holes. I'm shocked again to see them.

Our ground IV today is a veteran astronaut I've known for fifteen years, Megan McArthur. Despite being one of the youngest astronauts when she was selected, at twenty-eight, she's always been calm and sure of herself, even under pressure. She is talking us through our work today, and with her help, Kjell and I get ourselves and our tools out to the work site.

Our first task is a two-person job: removing a cover from a metal box and driving a bolt to open a valve that opens the flow of ammonia. Kjell and I get into an easy rhythm where it seems as though we can read each other's minds, and it feels as if Megan is right there in lockstep with us. We work together with an uncanny level of efficiency. With our visors almost pressed against each other, Kjell and I can't help but make occasional eye contact, and when we do, I get a sense he is thinking the same thing as me. Even though I'm not superstitious, I don't want to jinx it, or us, by saying, "This is going great" or "This is turning out to be pretty easy." We just need to keep it up until we are done.

When we get the cover back on the box, Kjell and I separate to work on different tasks for a while. He continues reconfiguring the ammonia lines, and I work on the vent lines on the back side of the space station's truss. Both are difficult tasks, and we are each absorbed in them completely. This is not the ammonia you might have found under your grandmother's sink, but something a hundred times stronger and much more lethal. If this ammonia were to get inside the station, we could all be dead within minutes. An ammonia leak is one of the emergencies we prepare for most. So working with the cooling system and the ammonia lines is especially important to get right the first time. We must make sure not to get any of the ammonia onto our suits.

As I had learned on my first spacewalk, I'm finding that the focus required to work outside is absolute. Every time I adjust my tethers, move a tool on my mini-workstation, or even just move, I have to concentrate with every bit of my attention, making sure I'm doing the right thing at the right time in the right way, double-checking that I'm not getting tangled up in my safety tether, floating away from structure, or losing my tools.

After a few hours, I head back toward the CETA cart (CETA stands for crew and equipment translation aid), which is sort of like one of those old manual handcars once used on railways. It's designed to let us move large equipment up and down the truss. When we were planning this spacewalk, I had raised concerns about whether this task, tying down the brake handle so no one could accidentally lock the brake, really needed to be done. This is much less important than our primary objective of reconfiguring the ammonia system, and it takes me far away from Kjell—too far to help if he runs into any trouble, like in my skydiving dream. The lead flight director insisted that we would be able to do both.

I'm plodding through my task with the brake handle, using reminders written on a checklist on my wrist. I am working mostly on my own, as Megan is concentrating on talking Kjell through his much more complex task. As I continue working, I can hear Kjell struggling with the ammonia connections. These can require all your strength, even for a strong guy like him, and they are technically complex, requiring upward of twenty steps each to mate or demate a connection, all the while remaining alert for ammonia to come shooting out and contaminate your suit. Each time I hear him struggle to complete a step, I again question to myself why I am working on the cart when I should be there to help him.

I finish up and take one last look over my work site, making sure everything looks right, before heading back out to the end of the truss to help Kjell. Hand over hand, it takes me a few minutes to get to him. I look over his suit, inspecting it for yellow spots of ammonia. I see a few places that look suspicious, but when I look closer I can see the threads of the suit material below the discoloration, which rules out ammonia as the cause. I'm glad I decided to wear my glasses, which haven't slipped or fogged up, or I might not have been able to tell the difference. We're preparing to vent the ammonia system—Kjell opens a valve and quickly moves clear. High-pressure ammonia streams out the back of the space station like a giant cloud of snow. As we watch, the sun catches the huge plume, its particles glistening against the blackness of space. It's a moment of unexpected beauty, and we float there for a minute, taking it all in.

When the venting seems to be complete, Megan instructs us to separate—Kjell will stay here and work on cleaning up the ammonia vent tool while I venture back to the solar array joint to remove and stow an ammonia jumper I installed earlier. The solar array joint continually rotates in the same direction to keep the solar arrays pointed at the sun, 360 degrees every ninety minutes, while passing electricity downstream. Megan talks me through the process. I struggle with one of the connections.

"Hey Megan. With the bale all the way aft, the white band should be visible or not?" I ask.

"Yes," Megan replies, "the forward white band should be visible."

I work with it for a few more minutes before getting it configured the way it's supposed to be.

"Okay," I report. "Forward white band visible."

"Okay, Scott, I copy the forward white band is visible—check the detent button is up."

"It's up."

When I hear Megan's voice again, there is a subtly different tone.

"I'm going to ask you to pause right here, and I'm going to tell you guys what we've got going on."

She doesn't say what this pause is about, but Kjell and I know: Megan has just been given some news within mission control, something the flight directors have to make a quick decision about. It may be something that puts us in danger. She doesn't leave us hanging for long.

"Okay. Currently, guys, from a momentum management perspective, we're getting close to a LOAC [loss of attitude control] condition," she says. She means the control moment gyroscopes, which control the station's attitude—our orientation in the sky—have become saturated by the venting ammonia. Soon we will lose control of our attitude, and when that happens, we will soon lose communication with the ground. This is a dangerous situation, just as we anticipated.

Megan continues. "So what we need Kjell to do is to pull out of your current activity and head over toward the radiator. We're going to have you redeploy it."

If we can't cinch down this radiator properly, we will have to put it back out in its extended position.

"Copy," Kjell answers crisply.

"You've probably gathered from a timeline perspective where we're going," Megan says. "We're going to have you clean up the vent tool eventually, Kjell. And Scott, you're going to continue with the jumper, but we are not going after cinching and shrouding the radiator today. It will take too long."

We both acknowledge her. This situation with the gyroscopes is serious enough to alter our plans. Even under the best of circumstances, when we hear we are close to saturating the gyros, it's one of those "Oh, shit" moments. The station won't start spinning out of control like a carnival ride, but losing communication with Megan and all the experts on the ground is never a good thing. And with the two of us outside, a communication blackout would add a new danger to an already risky situation. In all the preparation we've done for this spacewalk, we had never discussed the possibility of losing attitude control due to ammonia venting.

Houston is discussing handing over attitude control to the Russian segment. The Russian thrusters can control our attitude, less elegantly, with the use of propellant. The handover process isn't instantaneous, and we could lose communication with the ground in the meantime anyway. On top of that, the Russian thrusters use hypergolic fuel, which is incredibly toxic and a known carcinogen. If any of the hydrazine or dinitrogen tetroxide got onto our spacesuits, we could bring those chemicals back into the station with us.

But attitude control is important. If we can't talk to the ground, we lose the expertise of the thousands of people in Houston, Moscow, and other sites all over the world who understand every aspect of the systems keeping us alive up here. Our spacesuits, the life support systems within the station, the Soyuz meant to get us back safely to Earth, the science experiments that are the reason for us being here in the first place—our comm system is our only connection to the experts on all of these. Our only connection to Earth. We have no choice but to take the risk.

I think about just how alone Kjell and I are out here. The ground wants to help us, but we may not be able to hear them. Our crewmates inside the station would do anything to ensure our safety, but they can't reach us. Kjell and I have only each other. Our lives are in our own hands.

As instructed, we re-extend the spare radiator rather than taking the time to cinch it down and install a thermal cover. It will be safe in this configuration until a future spacewalk can retract it. We are nearing the seven-hour mark, the point where we were planning to head back to the airlock, but we are still far away with much left to do before we can get inside. We start the process of cleaning up our work site and inventorying our tool bags and mini-workstations to make sure we aren't leaving anything behind. Once everything is packed up and checked, we start the laborious process of traveling hand over hand back to where we started.

We are about halfway to the airlock when I hear Megan's voice again in my headset.

"Scott, if you're okay with it, we need you to go back to the vent valves and make sure they are in the right configuration. The specialists are seeing some data they aren't happy with."

This is a simple request, but Megan's tone communicates a lot—she wants me to know this action is not required and that I can say no without causing any problems. It's a task that could easily be left for the next astronauts, who will be launching next month. She knows that we have been outside a long time and are exhausted. My body is aching, my feet are cold, my knuckles are rubbed raw (some astronauts even lose fingernails from the intense pressure spacewalks put on our hands). I've been sweating and am dehydrated. There is still so much we have to do before we can get safely back inside, especially if anything unexpected happens between now and then.

I answer her right away, putting a vigor into my voice that I don't actually feel. "Sure, no problem," I say.

I've been convincing myself all day that I actually feel fine, that I have plenty of energy left. Both Kjell's life and my own depend on our ability to push past our limits. I've convinced myself so effectively that I've convinced the ground team too.

I head to the back side of the truss again to check the vent valves. It's dark now and starting to get cold. I don't waste the effort to adjust the cooling on my suit—even just that simple gesture would hurt my hands too much. I would rather just freeze.

In the darkness, I get turned around and upside down. I can see only what's immediately in front of my face, like a scuba diver in murky waters, and it's completely disorienting. Everything looks unfamiliar in the dark. (One difference between the Russian approach to spacewalking and ours is that the Russians stop working when it's dark; the cosmonauts just hang on to the side of the station and rest, waiting for the sun to come up again. This is safer in one sense—they are probably less likely to make mistakes, and to tire—but they also expend twice as many resources and do twice as many spacewalks because they only work half the time they are outside.)

I start to head in a direction I think is the right one, then realize it's wrong, but I can't tell whether I'm upside down or right side up. I read some mile markers—numbers attached to the handrails—to Megan, hoping she can help tell me where I am.

"It looks much different in the dark," I tell Megan.

"Roger that," she says.

"Did I not go far enough aft?" I ask. "Let me go back to my safety tether." I figure once I find the place where my tether is attached I'll be able to get my bearings.

"We're working on cuing up the sun for you," Megan jokes, "but it's going to be another five minutes."

I look in the direction I think is Earth, hoping to catch a glimpse of some city lights 250 miles below in the darkness to get my bearings. If I just knew which way Earth is, I could figure out where I am on the truss. When I look around, all I see is black. Maybe I'm looking right at the Earth and not seeing any lights because we're passing over the vast expanse of the Pacific Ocean, or perhaps I'm just looking at space.

I make my way back to where my tether is attached, but when I get there, I remember that Kjell had attached the tether, not me, so I'm not familiar with the area. I'm as disoriented as ever. I float for a minute, frustrated, thinking about what to do next.

"Scott, can you see the PMM?"

I can't, but I don't want to give up. I see a tether that I think is Kjell's—if it is, I might be able to figure out where I am.

"Scott," Megan says, "we're just going to send you back now—we don't need to get this, so just head back to where your tether location is and then head back to the airlock." She takes an upbeat tone, as if this is good news, but she knows it will be frustrating to me to hear they're giving up on me.

Eventually I catch a glimpse of lights above me. I'm not sure what it is at first, since above me is what I thought was the blackness of space. But as the lights come into focus, I see they are city lights—the unmistakable lights of the Middle East, Dubai, and Abu Dhabi, stretched along the Persian Gulf standing out against the blackness of the water and the desert sands.

The lights reorient me—what I'd thought was down is up—and I feel the strange sensation of my internal gyroscope righting itself. Suddenly it's clear where I am and where I need to go.

"I see the PMM now, so I think I'm close," I tell her. "I can go do it. I'd prefer to do it if you guys are okay with it."

A pause. I know Megan is consulting with the flight director about whether to let me continue or tell me to come back inside.

"Okay, Scott, we're going to take your lead. We're happy to have you go and do that."

"Okay. I think I'm in good shape now."

When I reach the work site, the sun finally shines over the horizon while Megan talks me through the steps of configuring the vent valve on the ammonia tank. Once I'm done, Megan tells us to head back to the airlock.

I contemplate making a joke to the ground by calling myself "Magellan"—a nickname we used to use in the Navy for those who got lost. But they might not get it, and besides, Magellan was killed before he made it home.

I head back to the airlock, where I climb in first this time and get my tether secured so Kjell can follow. He crams himself in behind me. As he struggles to close the hatch, I try to hook up the oxygen and cooling umbilical to my suit. But my hands are so fatigued I'm fumbling. To make matters worse, my glasses are positioned in such a way that I'm peering at the connection between the umbilical and my suit through the very bottom edge of the lenses, and the distortion prevents me from seeing clearly. I struggle for a good ten minutes, by which time Kjell has maneuvered himself into a position where he can see my connection and help me out. Working together, we get it connected. This is why we do spacewalks in pairs.

Kjell gets the hatch closed, and the air hisses in around us. The carbon dioxide in Kjell's suit is showing an elevated reading, so when the airlock finishes repressurizing, Kimiya and Sergey hurry to get him out of his helmet first. Through my visor, I can see that he is okay, nodding and talking. It will be ten minutes before Kimiya can take my helmet off. Kjell and I are attached to opposite walls, facing each other, held in place by the racks that secure our spacesuits. We have been in these suits for almost eleven hours. While I float there and wait to get out of my helmet, Kjell and I don't have to talk—we just share a look, the same look you'd give someone if you'd been riding down a familiar street together, chatting about this and that, and missed by nanoseconds being T-boned by an oncoming train. It's the look of realizing we've shared this experience that we both know was at the limit of our abilities and could have killed us.

When Kimiya lifts my helmet off my head, Kjell and I can finally see each other without plastic in between. We still don't feel any need for words.

Kjell smiles an exhausted smile and nods at me. His face is pale and sweaty, bathed in the artificial light.

Hours later, Kjell and I pass in the U.S. lab. "There ain't going to be no rematch," I say, quoting from *Rocky.*

"I don't want one," says Kjell, laughing his big boisterous laugh.

We have no way of knowing it yet, but only one of us is done with spacewalks.

. . .

A FEW DAYS LATER, I wake up to find a public affairs event on the schedule with the U.S. House of Representatives Committee on Science, Space, and Technology for both Kjell and me. We've been given no preparation or warning, as we should have been for such an important event, and because my day is packed between now and then, I won't have the chance to get ready. Even worse, when Kjell and I are connected we discover that we are being conferenced into a committee hearing, and that our participation will be considered testimony. I'm furious that the office of public affairs has not given me warning that I will be testifying before a committee of the people who oversee NASA and determine its funding. But I have to set aside that reaction and pretend to be prepared.

Kjell and I answer questions about what we're doing on station—we describe the biomedical experiments we are taking part in and talk about growing lettuce. One representative points out that we are in a "difficult geopolitical situation" with the Russians and wants to know whether we share all of our data with our Russian colleagues.

I explain that the international cooperation on the space station is its strength. "I was up here as the only American with two Russian guys for six weeks this summer," I tell him, "and if something had happened to me, I would have counted on them for my life. We have a great relationship, and I think the international aspect of this program has been one of its highlights."

One of the representatives on the committee, Dr. Brian Babin, happens to be a dentist, and the Johnson Space Center is in his congressional district. He is very curious about our oral health; we assure him that we brush and floss regularly. The last question is about Mars—a representative from Colorado points out that the planets will be lined up advantageously for a voyage in 2033. "Do you guys think that's feasible?" he asks.

I tell him I personally think it's feasible, and that the most difficult part of getting to Mars is the money. He knows what I mean without my having to spell it out—we can do it if his committee gives NASA the funding. “I think it’s a trip that is worth the investment,” I say. “I think there are things tangible and intangible we get from investing in spaceflight, and I think Mars is a great goal for us. And I definitely think it’s achievable.”

A COUPLE OF weeks later, we are eating breakfast when a fire alarm sounds. Even though we have had many false alarms over the time I’ve been here, this is still a sound that captures our full attention. It takes us only a few minutes to trace the alarm to the European module, where a biology experiment with rotating incubators is showing a slightly elevated carbon monoxide level. We power down the experiment and confirm that there are no elevated readings anywhere else in the station. We sample the experiment to see if there are signs of combustion. There are. It’s a real fire.

I always find the alarms entertaining in a strange way, unless they happen in the middle of the night and wake me up. Then I hate them. Alarms are a good reminder of the risk we live with and also a chance to review and practice our responses. In this case, the fire alarm does reveal an error in the procedure, which we later fix.

ON DECEMBER 6, a Cygnus resupply launches successfully. This is the first flight of an enhanced Cygnus with an extended pressurized module, which allows it to carry 25 percent more cargo. The module is named *Deke Slayton II* after one of the Mercury astronauts (the first Deke Slayton Cygnus blew up on launch the previous year). In addition to the regular supplies of food, clothes, oxygen, and other consumables, Cygnus is also carrying experiments and supplies to support research in biology, physics, medicine, and Earth science. It’s also carrying a microsatellite deployer and the first microsatellite to be deployed from ISS. And, important only to me, on board is a gorilla suit sent by my brother to replace the one lost on SpaceX. Once Cygnus safely reaches orbit—a stage we no longer take for granted after all the disasters earlier this year—Kjell captures it with the robot arm, his first time doing so. This was supposed to be my turn, but I decided to let Kjell do it, which meant giving up my last chance to grab a free-flying satellite, one of the few things I've never done in space.

A few days later, on December 11, we gather to say good-bye to Kjell, Kimiya, and Oleg. I remember when they arrived here about five months ago, which seems like another lifetime. Kjell and Kimiya, who showed up like helpless baby birds, are leaving as soaring eagles. They are now seasoned space flyers who can move around the station with ease, fix hardware of all kinds, run science experiments across multiple disciplines, and generally handle anything that comes their way without my help. It's a rare vantage point I have of their time here. I've known at some level how much astronauts learn and improve over the course of a single long-duration mission, but it's another thing entirely to witness it. I say good-bye knowing I still have three months ahead of me. I'll miss them.

YURI MALENCHENKO, Tim Kopra, and Tim Peake launch from Baikonur on December 15 at eleven a.m. our time and dock after a six-and-a-half-hour trip. I watch from the Cupola as they approach us, the bulbous black-and-white Soyuz with its solar panels spread like an insect's wings, a sight I never quite get used to. The capsule starts out so tiny it looks like a toy, like a scale model of itself, at times appearing to be on fire as the sunlight reflects off its surface. But then it gets bigger and bigger, slowly revealing itself to be a full-size spacecraft.

As I watch it get closer, I start to feel that something about the approach isn't quite right, the angle or speed or both. The Soyuz is too far forward from its docking port. Just a few meters away from the station, the Soyuz stops by firing its braking thrusters toward us, to hold its position. This is not normal.

The Soyuz thrusters are blowing unburned propellant at the Cupola windows, so I hurry to close the shutters. Beads of the stuff bounce around between the window and the shutters, a strange and alarming sight. I rush down to the Russian service module to find out from Sergey and Misha what's going on.

"They had failure in automatic docking system," Sergey tells me. No one knows why. Yuri, the Soyuz commander, takes over manual control, and after spending a few minutes getting realigned with the docking port, he guides the craft in successfully, just nine minutes behind schedule. This is a great example of why we train so much for things that are unlikely to happen. The automated system is generally reliable, but a failure could cost the crew their lives if someone isn't ready to take over.

After a leak check, which always seems to take longer than it should—a couple of hours this time—we open the hatch and welcome our three new crewmates aboard. As always, their first day is a full one. Throughout, I'm aware that this is the last time I will introduce new people to space, and it gives me a strangely sad feeling, a kind of pre-nostalgia.

I don't know Yuri especially well, though he is one of the most experienced space travelers in history. He has a reputation for being technically brilliant, and his handling of this emergency manual docking will likely only add to that reputation. He has flown in space five times, including a long-duration mission on *Mir,* a space shuttle mission, and three previous long-duration missions on the International Space Station for a total of 641 days in space. He also has the distinction of being the only person to have gotten married while in space—on his first ISS mission, in 2003, he and his bride, Ekaterina, exchanged vows via videoconference while her friends and family gathered around her at home in Houston. (Knowing Yuri, I'm pretty sure he wasn't crazy about this idea, but he went along with it anyway.) On his fourth flight, in 2008, Yuri's Soyuz landed so far from his intended touchdown point, the local Kazakh farmers who came upon his steaming spacecraft had no idea what it was. When he and his two female crewmates, Peggy Whitson and Yi So-yeon, emerged from the capsule, the Kazakhs mistook him for an alien god who had come from space with his own supply of women. Had the rescue forces not arrived, I suspect the farmers would have appointed him their leader.

Tim Kopra was an Army aviator and engineer before joining NASA in the 2000 class. He went to West Point and is a colonel in the Army.

He also has multiple master's degrees—one in aerospace engineering from Georgia Tech, one in strategic studies from the Army War College, and one in business administration from a joint program between Columbia University and the London Business School. He's been an astronaut for fifteen years, but this is only his second time in space. He flew one mission on ISS in 2009 that was unusually short—just more than a month. He had been scheduled for a second mission in 2011 on the space shuttle, but a few weeks before the launch he fell off his bike and broke his hip.

Tim Peake was a helicopter test pilot in the United Kingdom until he became the first official British astronaut chosen by the European Space Agency. This is his maiden voyage to space, making him the only rookie on the crew. For the UK, Tim is sort of their Yuri Gagarin and Alan Shepard rolled into one. That's a lot to live up to, but as our mission together goes on, I will come to find that he is more than up to it.

One of the first things Tim Peake does when he gets on board is to open up a package of bonus food that came with him, select a BLT sandwich (a "bacon sarnie" in British English), and eat it, pieces of bacon floating out tantalizingly in every direction. Tim doesn't realize that this sandwich, from the European Space Agency, is not available to the rest of us. We haven't had a proper sandwich in months (for me, nine months), so watching him eat it is a very special form of torture. When he notices us eyeballing his sarnie, Tim offers Misha and me bites. Afterward, we watch him finish eating, salivating like two dogs on steak night.

As always with newcomers to the space station, Tim and Tim are as awkward and clumsy as toddlers. Sometimes when I want to help one of them get to where he needs to be, or want to get one of them out of the way quickly, I find it's easiest to physically grab him by the shoulders or the hips and move him around in space the way I'd move a bulky piece of cargo. Neither of them seems to mind.

The next day, I hear during the daily planning conference that we have a problem. The mobile transporter is stuck—it is attached to the CETA cart I worked on during my second spacewalk with Kjell. Flight controllers had started to move it to a different work site near the center of the truss so the robotic arm would be in position to perform some maintenance activities before the new Progress arrives next week. But it quickly became stuck in a position that makes dockings impossible for future visiting vehicles. The moment I hear these words, my heart sinks. I immediately know what went wrong: when I was working on the CETA cart, I must have inadvertently locked the brakes while tying down the brake handles.

"I think I know who screwed that up," I tell Houston.

Later, I get on the phone with the flight director and tell her I'm almost certain about the brake handle.

There is a pause on the other end of the connection. "How certain are you?" she asks.

"Very certain," I say. I know what my answer will mean: I will have to do an unplanned spacewalk before the Progress can dock, and it's launching a week from tomorrow. We'll have a terrifyingly short period of time to prepare, both in space and on Earth.

It's important to me to admit mistakes immediately, and I don't make excuses. But I think to myself that I should not have been asked to do that task in the first place when I couldn't give it the proper amount of attention. Working with the cart had been an afterthought, and there is no place for afterthoughts when you're outside in space.

If some other piece of equipment had been left in the wrong configuration, we could likely wait until the next scheduled spacewalk, even if it was months away, to fix it. But the mobile transporter is now stuck between work sites and isn't secure enough to withstand the forces of Progress dockings. In addition to making it impossible for us to dock visiting vehicles, the stuck cart prevents us from moving the station in order to avoid debris, firing thrusters to reduce the momentum on our gyros, or using the robot arm for anything else. I start mentally preparing myself to go outside for the third time. I share the news with the Russian crew, and they say they will help in any way necessary. The next day, NASA makes the official decision that we will try to fix the transporter on an emergency spacewalk.

It's hard enough to prepare for a spacewalk under ideal circumstances; it's much harder to do it on short notice and with colleagues who are still acclimating to this strange environment. Tim Kopra, though an experienced astronaut, has been here only a few days and is still adjusting to living in space. He will have to get into a spacesuit and venture outside with me. Tim Peake, who is still figuring out the most basic aspects of life up here, like eating and sleeping, will have to serve as IV. Both of them will have little margin for error in their demanding jobs.

I email Amiko that I may have to go outside again next week, expressing my frustration with myself for being such a dumbass to have locked the brakes. She is sympathetic—she knows better than anyone but Kjell what the previous two spacewalks were like both physically and mentally. I also mention to her that Tim Kopra has a strange habit of repeating me. If I say, "I wonder if there's any football on today," Tim will say, as if I had never spoken, "I wonder if there's any football on today." I mentioned to Tim that my calf muscles have shrunk dramatically since I've been here, and he immediately responded, "My calves have shrunk dramatically too."

"But Tim," I said, "you just got here."

"Yeah, but I have really small calves."

This is not something I ever noticed about Tim working with him on the ground, and it's not like me to get annoyed by a crewmate—I've made it this long without being annoyed by anyone, which I think proves I'm pretty tolerant. Amiko suggests that maybe Tim is feeling insecure about joining me as a crewmate when I've been up here for so long. I agree and tell her I also wonder if it's just me.

The Russians spend the next few days packing trash into the Progress, which will be departing soon to burn up in the atmosphere. They have some extra room and ask if we want to put trash on board. Like many things in space—oxygen, water, food—trash removal capacity is a resource, and our two countries trade it like currency. I give them a couple of our large trash bags without telling Houston I did so. It would create a lot of work for people on the ground to ask permission, and we would likely not be granted permission anyway. I've been doing this all year, sneaking trash off the space station when the Russians have room for it, and we help them out too when we can. (This will cause a problem later when we pack the Cygnus and don't have all the trash Houston thinks we should—ten bags. After a lot of questions, I eventually tell them, "The trash fairy must have come in the middle of the night." No one mentions it again, which is a relief.) On December 19, I watch from the Russian service module as Sergey and Yuri monitor the departure of Progress on their displays as it inches away almost imperceptibly. As with the Soyuz, they can take over manually if there is a malfunction, but everything goes according to plan. Now that Progress is gone, we have room for the new Progress that will launch in a few days. I realize the next time something pushes away from the station, two and a half months from now, it will be me.

IN THE MORNING, I find an email from the ground asking me to submit a guest list for my landing. A limited number of people will be allowed to come to the control center in Houston to watch on the big screens as our Soyuz lands in Kazakhstan. I start making a list: Amiko, Samantha, and Charlotte. My dad, Mark, and Gabby. Gabby's chief of staff, Pia. Amiko's sons, Corbin and Tristan. My friends Tilman, Todd, Robert, Gerry, and Alan. Sarah Brightman. I picture the spectator area in mission control, canted seats behind glass, my friends and family gathered there and watching as our capsule falls through the atmosphere and lands—we hope—safely on the desert steppes of Kazakhstan.

Suddenly it occurs to me that making this list is the first thing I am doing to prepare for my return to Earth. From now on I will do more and more—throwing things out, packing things up, making more lists, thinking about what my next steps will be in life. I have a lot more time in space to go, but as of today a small part of my mind is on my future on Earth.

Tim, Tim, and Yuri will still be up here when I leave, less than three months from now. As the end is drawing nearer, my remaining time seems to stretch out longer, like taffy. I'm three-quarters of the way through—it should all be downhill from here. Yet when I allow myself to think about it, I remember the first three months, the way it felt when my first set of crewmates left, how long it felt like I'd already been here, how long ago that was. I can barely remember what Terry, Samantha, and Anton look like, what their mannerisms are, or what their voices sound like, the sound of Samantha humming. Like old friends who drifted away long ago, they are now a distant memory.

Running through the rain, driving a car, sitting outside, smelling fresh-cut grass, relaxing with Amiko, hugging my kids, deciding what to wear—it is hard to remember such acts with any specificity. I no longer have any sensory reminders of what they feel like. I am a fully acclimated space creature, and my return doesn't seem much closer than it was when I started. I will still be up here, in the same small spaces, for months.

One day soon after, I'm answering some emails and come across an invitation to speak at a conference in April. When I open my calendar, I realize that I'm scheduling my first event for after I get back to Earth.

ON MONDAY, December 21, I wake up early, diaper up, and get into my liquid cooling garment for the third time. Tim Kopra and I start our prebreathe of pure oxygen, then an hour later Tim Peake helps us get into our spacesuits. This spacewalk will be shorter than the previous ones—we will get the CETA cart and the mobile transport unstuck, then do a couple more tasks we know will need to be done at some point (called get-ahead tasks) so as to make the best use of the time and resources it takes just to get suited up and get out the door. Tim Peake does a great job as IV—as he moves through the checklist to get us ready (with the help of Sergey), any concern I might have had about whether he was prepared to take on this role after only being up here for six days dissolves. He works efficiently and confidently, and soon we are in the airlock and doing our leak checks.

I'm wearing the spacesuit with red stripes again, EV1. When the airlock is fully depressurized, Tim Kopra and I switch our suits to battery power and the spacewalk has officially begun. This is Tim's second spacewalk, but his first was in 2009, so it's been a while. Once we are outside and have completed our buddy checks, I translate to the CETA cart. When I reach the cart, I try moving it along the truss and, sure enough, it's stuck. I release the brake handle, then move it freely in both directions. The ground is satisfied.

It feels odd to have accomplished our main objective only forty-five minutes in. We finish up some of the tasks that Kjell and I had to leave undone the last time we were out here, mostly routing cables to locations where they can later be connected with future hardware. We come back inside after three hours and fifteen minutes, and while I'm far from the exhaustion I felt at the end of my earlier spacewalks, I'm still tired and sore. My fatigue level is more like what I used to feel after the training runs we did in the pool in Houston. Much easier, but still not easy.

After I come back in, I speak with Amiko and then check my email. There is one from Kjell, telling me he had watched the spacewalk on NASA TV. It's strange to imagine him in Houston, watching in the predawn early morning, sitting in a chair of some kind, gravity holding him there. "You guys crushed it!" his email reads. He asks about what we did and what the experience was like with the specificity and enthusiasm of someone who had just been there. In my response I tell him it was less than half as long as our shorter spacewalk but required only about a fifth the effort. I tell him either the time versus perceived effort is exponential or else those were just exceptionally tough spacewalks he and I did.

"How's Earth?" I type to end my message. "Starting to forget what space is like yet? Merry Christmas!"

For the rest of my mission, I occasionally look out the window and catch sight of the area at the end of the truss where Kjell and I worked on that second spacewalk. It looks far away, farther than home, and it gives me a strange feeling of nostalgia, like the feeling I get when I visit my old neighborhood in New Jersey. Not just a place I've spent time, but a place imbued with strong emotions, a place familiar yet at the same time distant, now unreachable.

18

December 24, 2015

Dreamed I met with General David Petraeus and he was trying to warn me of something. Some kind of trouble I would experience on this flight. Then I was on a U.S. aircraft carrier in the Indian Ocean off of Oman. We heard there was a hurricane coming with two-hundred-mile-per-hour winds, and soon after it came out of nowhere and capsized the aircraft carrier. Then the crew rebelled against the officers.

TODAY IS Christmas Eve, my third in space. This isn't a number anyone would envy, especially a parent with kids: a holiday that celebrates family togetherness can be the toughest time to be away. On top of that, the last two weeks have overextended us. The previous crew leaving, the new guys arriving, helping them get acclimated, preparing for and executing the emergency spacewalk—these have all been demanding, and they have come one right after the other. I have worked nearly two weeks without a day off, so my mood going into Christmas is less than festive.

Holiday or not, today is just another workday on the schedule, one that becomes more difficult when the resistive exercise device breaks down. This is more urgent than it might seem, because exercise is nearly as important to our well-being as oxygen and food. When we skip even one exercise session, we can feel it physically, as if our muscles are atrophying, and it's not a good feeling. Tim Kopra and I take nearly half the day to fix the machine—a broken damper, like a shock absorber, is the culprit. Because of this, we don't wrap up our workday until eight p.m., by which time I've missed both my exercise sessions, which only adds to my crappy mood.

I call Amiko before heading over to the Russian segment for dinner. I've felt like something has been bothering her for the past few days, and I'm starting to get the sense that I've done something to upset her. Maybe it has taken me longer than it should have to figure this out, since it doesn't seem like there is much I can do to annoy her from space.

I reach her on a checkout line at the grocery store. Not ideal for an emotionally honest conversation, but we don't have much time left on this comm pass, so we have no choice.

"I get the sense that something has maybe been bothering you," I say. "Have I done something to upset you?"

She thinks for a moment, then gives a long sigh. She sounds exhausted.

"I feel like when you get back we will have to reconnect," she says.

Of course we will have to reconnect, I think—I will have been away for a year. "What does that mean?" I ask. "You feel—disconnected?"

Amiko explains that the holidays are tough because she is missing not only me but my daughters. She is carrying a heavy load watching over my father and her own sons, taking care of our house and the many things I can't be there to help her with. Her already demanding job has become more stressful—she is being edged out of her social media management position, and her supervisors have made clear that she cannot help me with my social media on work time—a counterproductive policy when I have over a million Twitter followers and similar numbers on other social media platforms. She is forced to use her own time or take annual leave to conduct interviews about my mission or even just to walk over to the astronaut office to drop off items to be included in my care packages. All of her hard work and sacrifice go completely unrecognized by her management. (By contrast, my colleagues in the astronaut office have been unfailingly supportive of Amiko as my partner, for which I've been grateful.)

The strain of all these pressures has been taking a toll, and Amiko has hidden it from me. She enjoys working with me on my social media campaign and takes pride in how successful it's been, but lately many of our conversations revolve around things I need her to do to the exclusion of everything else. At times she can feel like my coworker rather than my partner. Worse, she tells me it's starting to bother her that she no longer remembers how I feel or smell, what it's actually like to be with me face-to-face. She says she is craving real human touch. Then the satellite drops out and I lose her in midsentence.

I float by myself in my crew quarters for a few minutes, knowing I won't be able to get her back on the phone for a long while. I have full confidence in our relationship, and Amiko has been nothing but loyal and honest in the six years we've been together. But hearing the words "reconnect" and "real human touch" feels simply awful. Amiko is attractive, and she would have no problem fulfilling any cravings for human contact she might have on Earth. I'm not the type of person to get jealous, and jealousy isn't exactly what I'm feeling. It's more like I'm letting my imagination run wild while orbiting the planet, as physically far away from Amiko as I can be, and letting the reality of the situation sink in. She wants something very simple, and I can't give it to her.

I make my way to the Russian segment and put a fake smile on my face. The Russians don't celebrate Christmas at the same time we do—the Orthodox calendar has Christmas on January 7—but they are happy to host a festive meal for the rest of us nonetheless. I discover that the nutritionists in charge of our food have not bothered to create a special holiday meal, so I eat turkey cold cuts doused in a salty brine as Christmas Eve dinner. We do, however, have some hard salami that came up on Cygnus and some of that black, tarry caviar from the Russians, as well as some fresh onions and apples that came up on Progress yesterday. Many toasts are made by all. We listen to Christmas music and the new Coldplay album I was recently uplinked, which everyone enjoys. We toast our privileged spot in space, how lucky we are to be here and how much it means to us. We toast our family and friends back on Earth. We toast one another, our crewmates, the only six people off the planet for Christmas.

An hour and a half later, I get my scheduled videoconference with my kids. Samantha has traveled from Houston to Virginia Beach to be with her mother and sister for the holiday, and I'm pleased to see my girls together. They seem happy to see me, though they also seem uncomfortable. From what I can tell, the apartment doesn't seem very Christmassy, and I hope the girls are having a better holiday than I am.

Later I'm able to get Amiko on the phone again, and she tells me something she has never told me before—that because I continue to make an effort to make this year in space look easy, it can seem as though I don't miss her and don't need her. We both take pride in being strong and making difficult things seem easy. But by keeping the strain to myself, I shut her out. I tell her that making it look easy is the only way I can convince myself I can do this, but in reality it isn't easy at all. I have figured something out recently: Amiko has only me to miss, and all the other aspects of her life are more or less unchanged. I have her to miss, but also my daughters, my brother and father, my friends, my home, showers, food, weather—literally everything on Earth. Sometimes my missing her can be obscured by how much I miss everything, and I can see how this would make her feel that she is somehow more alone in this than I am. And she is right.

I don't get much sleep, and in the morning I float awake in my sleeping bag, putting off starting my day. Christmas mornings when I was growing up in New Jersey, my brother and I used to leap up even before it was fully light and run to the living room in our underwear to find our presents. When they were little, my daughters did the same. Later today I will do some public affairs events, and I will be asked what it's like to spend Christmas in space. I will answer that being here at this special time gives me a chance to reflect on the holiday and how lucky we are to be able to see this view of our planet. I will say that I miss being away from the people I love. For now, I'm just floating here while my crewmates are still asleep, a glowing computer in front of me and the fan humming loudly beside my head.

. . .

As we get toward the end of long-duration missions, our trainers at Johnson Space Center start to slowly ramp up our resistance exercise in order to acclimate our bodies to the stresses of being back in gravity. I remember this from my previous mission—and I remember not enjoying it—and though I understand the necessity I'm also concerned about injuring myself. If I had a serious injury and couldn't exercise, it would make life far more difficult when I get back to Earth's gravity. The next afternoon, I'm doing squats with a heavy load when I feel a searing pain in the back of my leg. It doesn't take me long to realize what's happened: I've torn a muscle in my hamstring. The pain doesn't go away, and now I can't work out.

My flight surgeon, Steve, prescribes the muscle relaxant Ativan. We have a stash of drugs—including Ativan and many others—secured in a bag on the floor of the lab module with our other medical hardware. The bag contains medications of all types: painkillers, antibiotics, antipsychotics, just about anything you would be able to find in a hospital emergency room. The controlled substances have warning labels from the DEA authorizing access only under doctors' orders. NASA plans for everything—we even have an early pregnancy test and a body bag.

The next morning, I email Amiko a picture of an orbital sunrise. Because we use Greenwich Mean Time on station, I have a five-hour head start, and I know this will be waiting for her when she wakes up. I tell her that this photo is not for social media, but just for her. Later she tells me this is just the sort of virtual hug she has been hoping for. I can't be there to make things easier for her, but I can at least let her know I'm thinking of her.

In the afternoon, I'm preparing some lunch when Tim Kopra floats by looking for something to eat.

"This chicken soup is really good," I tell him.

"The chicken soup is really good," he says as if I'd never spoken.

"Yup. I'm also going to have some of that barbecue beef," I say. We watch CNN together for a few minutes while eating.

After a bit I say, "You know, on second thought, I don't like this soup."

"Yeah, I don't like it either," Tim says. When we finish our food, we each get back to our respective tasks. It takes me a few minutes before I realize I'm not annoyed by Tim's repeating what I just said. It also doesn't bother me when we lose the satellite signal and the story I'm following on CNN cuts out. It doesn't even bother me when a tiny brown sphere of barbecue sauce propels itself onto the thigh of my pants. I feel calmer and more content with my surroundings than I have in months, maybe all year.

That evening, I tell Amiko about this strange effect of the muscle relaxant.

"You're under a lot of stress," she points out. "The drug will affect that."

I tell her my flight surgeon mentioned that the drug is sometimes prescribed for mood and anxiety disorders. "I haven't *felt* that stressed out," I tell her. In fact, I've felt pretty normal, all things considered. But I guess just being here has been getting to me. I have to set aside stress so I can concentrate on what I need to do, but when stress is always there, it can come out in unexpected ways—like feeling annoyed by a colleague. I also have to keep in mind that I've been living with high CO_2 for almost a year, which is known to cause irritability. At any rate, it's nice to feel better, and I try to enjoy the positive side effect of the drug while it lasts.

That night, I read a few pages of the Shackleton book in my sleeping bag. On Christmas 1914, the first officer of the expedition wrote in his journal, "Here endeth another Christmas Day. I wonder how and under what circumstances our next one will be spent. Temperature 30 degrees." He couldn't have imagined how he would spend his next Christmas—camping on an ice floe with minimal provisions after their ship, the *Endurance,* had been crushed by ice. For all the suffering of their ordeal, the men discovered they enjoyed the self-reliance they had found. "In some ways they had come to know themselves better," author Alfred Lansing writes. "In this lonely world of ice and emptiness, they had achieved at least a limited kind of contentment. They had been tested and found not wanting."

I turn out the light and float for a while before falling asleep.

NEW YEAR'S EVE IS a bigger holiday than Christmas on the space station because it's celebrated by all nations on the same day. We gather in the Russian segment for the festivities. We all have something to eat, someone makes a toast. We continue that way on into the night. We briefly turn out the lights to see whether we can glimpse any fireworks on Earth—on my previous long-duration flight we were able to see the tiny specks of colored light, but this year we don't see any. It is still a privilege to be spending my second New Year's Eve in space, and I'm glad I'm still able to appreciate where I am and what I am getting to do. The next morning I get up early to call my friends and family in the United States to wish them a happy 2016—the year I will come home.

19

January 7, 2016

Dreamed Amiko and some of my astronaut colleagues arrived unexpectedly on the space station. They had come on a bus, which made sense in the dream. I went to the U.S. lab to clean up a bit and found a cigarette my dad had left smoldering. The cigarette started to catch on loose papers floating around. I shouted at everyone to evacuate, and I stayed behind to fight the fire with a garden hose, which I was surprised to find coiled on the wall along with all of the other equipment we keep in there. It didn't work very well, though, because the space station was made of dried wood. The fire grew until there were flames all around me, and I fought it until I woke up.

Tomorrow is the fifth anniversary of the shooting in Tucson that injured Gabby. The approach of this day has made me think about what I was doing when Gabby was shot. I was fixing the toilet, and now the same toilet is failing in the exact same way.

As weird as this is, we've sort of known this day was coming. We were hundreds of days beyond what was supposed to be the lifespan of the toilet's pump separator, and I had started to see it as a challenge to keep it limping along past the point when the ground had suggested I should replace it. I should have listened to them, because if it fails catastrophically it produces an enormous sphere of urine mixed with the sulfuric acid pretreat, a gallon of nastiness I would have to clean up.

Five years ago, I was orbiting 250 miles above my family when they needed me most. So much has changed since then; yet I'm in the same place, doing the same thing, as we honor the victims of the shooting with a moment of silence.

TODAY, January 15, 2016, is a great day on the International Space Station because a spacewalk is going on and I'm not doing it. I will always be glad that I had the chance to experience the thrill of floating outside the station with nothing but a spacesuit between me and the cosmos, but at least for now, I am more than fine with sitting this one out. It's also a thrill to see Tim Peake become the first official British astronaut to do a spacewalk.

Today I am serving as IV. I make sure both Tims get into their suits properly, call out the steps as they check over their tools and check their spacesuits' functions, and operate the airlock. Tim and Tim are replacing the power regulator that failed back in September while Amiko was on console in mission control, as well as installing some new cabling. They get through those tasks and a few others successfully before Tim Kopra's CO_2 sensor malfunctions. This isn't a big deal in itself, because he can self-monitor his CO_2 levels based on his symptoms, but soon after, he reports a water bubble inside his helmet. If the bubble were small, we might have speculated that it was a drop of sweat that had broken free, but the bubble is big. Tim also reports that when he pushes his head back against the absorbent pad in his helmet, there is a squishing noise, a sign that more water than just that bubble is in his helmet.

Tim and Tim have been outside for only four hours, and they have a number of items left on their to-do list, but a water leak in a helmet always means it's time to come back inside, *now.* Tim Peake will clean up the work sites as Tim Kopra heads immediately back to the airlock. We want to get them back inside quickly, but rushing increases the chances of something going wrong. So we go through the procedures methodically, one by one, so as to be sure we aren't screwing anything up. I think of a saying I once heard that was attributed to the Navy SEALs: "Slow is efficient. Efficient is fast. Slow is fast." When I bring them back inside, I get Kopra's helmet off first. He seems fine, if a bit moist. Then we get Peake's helmet off. They both look tired, but neither of them have the exhausted look that Kjell and I had after our first two spacewalks. We'd been in our suits almost twice as long.

A few days later, the Seedra in Node 3 fails again. Often when it goes down, the ground can get it up and running again, and I spend the day hoping they will be successful. I've been daring to hope that I could get back to Earth without having to mess with the Seedra again, so when the ground tells us that Tim Kopra and I will have to take it apart and spend a couple of days on repairs to get it working, I acknowledge it with a heavy heart.

The next day, Tim and I slide the beast out of its rack, move it to Node 2, secure it to a workbench, and take it apart. Over the course of the day, we isolate the problem. When I took this machine apart with Terry Virts, it was a multiday operation that left him bandaged and both of us tired and frustrated. Today, I'm aware from the start that the repair is going much better than it had previously. It's still a very complicated and challenging job—just moving the five-hundred-pound mass is a hassle when it could damage a hatch seal, sensitive equipment, or a body part. But I've had so much experience with this machine that I can now work on it with an incredible level of confidence and efficiency. At this point, I could write a repair manual for this damn thing if I wanted to. I feel like I know it the way a cardiologist knows a human heart.

We save time by using tricks I've figured out on previous repairs and get the work done in a fraction of the time it took Terry and me to do it back in April. I can't help but take some pride in that. I also can't help but wish fervently that I will never have to take this thing apart again.

Later in the day, I'm working in the Japanese module when I come across a drink bag wedged behind a piece of equipment. I dislodge it and find it's marked with the initials DP. No one up here has the initials DP, or has in a long time. It must have belonged to Don Pettit, who was last here in 2012. I save the bag until Don is working as capcom, then hold it up in front of the camera and ask, "Is this your drink bag?"

Don laughs at the absurdity of the situation. But he understands, as all space station astronauts do, how easily objects get lost up here. At home you would never put down a glass of water and lose it for three years, but up here, as careful as we are, it's incredibly easy to lose your drink or anything else. There is just too much stuff, and it all floats.

A few days later, I take a great picture of the city of Houston and the Gulf Coast on a beautiful night pass. When I send it to Amiko, I use the word "home," and I'm surprised to find that I'm starting to set my internal compass there again. I'm starting to allow myself to look forward to getting back. I hadn't been able to indulge in these kinds of thoughts for most of the year, but now it actually feels good to yearn for home a bit, knowing I'll be there soon.

Later in January we see through the second major botanical project on board the station. Growing lettuce back in August was relatively easy—we set up nutrient "pillows" under the grow light in the European module, watered the plants according to schedule, and watched the leaves sprout as expected for an easy harvest. Now I am growing flowering plants, zinnias, which we expect to be more difficult because the plants are more delicate and less forgiving. The sequence was set up this way on purpose—we will use what we learned from growing an easier, less demanding species to aid us in growing something more finicky. The zinnias prove to be even more difficult than we expected. They often look unhealthy, and I suspect that our communication lag between space and the ground is to blame. I take pictures of the plants and send them to scientists on Earth, who, after looking them over and consulting among themselves, send me instructions about what to do—usually "water them" or "don't water them." But the lag in communications means that by the time I get the instructions things have gone too far in one direction or the other. By the time I'm told not to water them, the little plants are often waterlogged and growing mold on their leaves and roots. By the time the instruction to water them reaches me, they are dehydrated and on the brink of dying. It's frustrating to be growing a living thing up here and to watch it struggle, not to be able to take proper care of it. At one point I post a picture of one of the zinnias on social media and get back criticism of my botany skills in return. “You’re no Mark Watney,” quips one smart-ass commenter, making reference to the stranded astronaut in *The Martian*. Now it’s personal.

I tell the payload ops director that I want to take over deciding when to water the flowers. That might seem like a small decision, but for NASA it’s huge. Having to touch the plants and the medium they grow in with my bare hands would be a major change in protocol. The ground seems terrified that if I touch the plants and they have mold on them, the spores could infect me. The initial reaction I get is skeptical, but I’m convinced that the flowers are going to die unless I’m allowed to take care of them myself, as a gardener on Earth would do, and it’s frustrating to see all the effort and expense that went into designing and launching this experiment going to waste. Some involved with this decision doubt I will check on the plants every day, because that will take a lot more of my time and attention than simply following directions. But I finally get my way.

It’s hard to describe the feeling of watching the flowers come back from the brink of death. I’ve carried memories of the flowers I saw in the botanical gardens with my grandparents when I was a kid, and maybe because those weekends with them were a peaceful respite, I associate flowers with my grandmother and her loving manner. I think about Laurel’s violets that I kept in my office after her death. Once the zinnias are my personal project, it becomes incredibly important to me that they do well. I check on them as often as I can. One Friday I bring some of them down to the Russian segment and attach them to the table as a centerpiece.

“Scott,” Sergey says, a puzzled look on his face. “Why are you growing these flowers?”

“They’re zinnias,” I clarify.

“Why are you growing these zinnias?”

I explain that we are working toward being able to grow tomatoes one day, that this is one of the experiments we are doing to further our knowledge for long-term spaceflight. If a crew is going to go to Mars, they will want fresh food and won’t have access to resupply like we have on the space station. If we can grow lettuce, maybe we can grow zinnias. If we can grow zinnias, maybe we can grow tomatoes, and tomatoes would provide real nutritional value to Mars travelers.

Sergey shakes his head. "Growing tomatoes is a waste. If you want to grow something you can eat, you should grow potatoes. You can live on potatoes." (And make vodka.) The practical and simple Russian perspective has merit.

When I post the first picture of the healthy zinnias on social media, there is a huge explosion of interest—6 million impressions. It's gratifying to see people respond with enthusiasm to something I've come to care about. And it reinforces my thinking that people are interested in what's going on in space if it's presented to them in ways they can relate to.

To me, the success with the zinnias is a great example of how crew members will have to be able to work autonomously if we ever go to Mars. I care about the flowers much more than I was expecting to, partly because I've been missing the beauty and fragility of living things, but most likely because I was called out on Twitter for my botany skills and had something to prove.

IN LATE JANUARY, I gorilla up for the first time, stuffing my head and body into the plasticky-smelling gorilla suit. I've decided what Space Gorilla's first adventure will be: I hide in Tim Kopra's crew quarters and wait for him to come along. When he opens the door, I pop out and scare the shit out of him. Then I float down to the Russian segment and show the cosmonauts, who all go nuts laughing. Space Gorilla is already spreading joy.

I decide it will be funny to float in front of the camera where mission control can see me in the suit without warning them first. On a calm Tuesday afternoon when not much is going on, I make my move. I put on the suit and then drift in front of the camera in the U.S. lab until I know I can be seen on the screen. Amiko sees it on NASA TV, but no one on the ground says anything. It's a letdown.

I've been thinking about ways to use Space Gorilla to engage with kids—if he could grab their attention and make them laugh, maybe they would be interested enough to listen to me talk about space and the value of science, technology, engineering, and math. Tim Peake agrees to help me by costarring in a short video in which he is shown unpacking some cargo, only to find a stowaway gorilla that chases him up and down the U.S. lab to the *Benny Hill* theme song, "Yakety Sax." The video goes viral and brings new attention to what we are doing on the space station.

On January 28, I lead a moment of silence for the crew of the space shuttle *Challenger,* which was lost thirty years ago today. The two Tims and I gather in the U.S. lab, where I say a few words honoring the memory of the crew and mentioning that their spirit lives on in our current achievements in space. I bow my head for a moment, and as I do I can't help but remember the cold morning when my college roommate George and I watched the orbiter blow up over and over on his tiny TV. Thirty years, a lifetime ago. I couldn't have imagined where I am now. I remember George asking me whether I still wanted to go to space and wanting to go more than ever.

A few days later, one of my Russian colleagues floats over to the U.S. segment to show me that his tooth has popped out. It's a crown attached to an implant, like a little metal peg in the front of his mouth. There is no way to get home without the tooth-jarring Soyuz landing, so he is understandably afraid that the unsecured tooth will be knocked down his throat, or lost, on reentry; he also doesn't want to land without it, because we are photographed so much immediately on return. I get out the dental kit, thoroughly dry both the tooth and the post with gauze, mix up some dental cement, and glue the tooth back into his mouth. My colleague gives me a broad smile, satisfied. A commander's work is never done.

One Sunday morning, I float over to the Russian segment and greet the cosmonauts while they are having breakfast.

"Scott!" Misha calls out to me with a mischievous smile on his face. "Do you know what today is?"

"Yes," I answer. "It's my birthday. February twenty-first." The last time I celebrated at home in Houston, I was fifty. Today I'm fifty-two.

"Happy birthday, Scott, but that's not it! We have only nine days left!"

I've avoided counting down this whole year. I'm surprised that the single digits have crept up on me, so it seems my strategy has worked. Nine days isn't long at all.

"Scott," Misha says with a note of excitement in his voice, "we did it!"

"Misha," I answer, "we had no choice!"

Sergey, Misha, and I will do a few Soyuz training sessions together so we will be ready for our descent. Misha, who will serve as flight engineer 1, needs to refresh his training for serving as Sergey's backup; it's been a long time.

We start packing up our things and getting organized to leave. I have to figure out what is coming back with me on Soyuz—a small package of no more than a pound or so, including the gold pendants for Amiko, Samantha, and Charlotte and silver versions for my crew secretary Brooke Heathman, my scheduler Jennifer James, and my Russian instructor Elena Hansen. A larger allotment of things can come back on SpaceX later in the spring. I need to clean out my crew quarters thoroughly so it will be fresh for the next person. Because of the way stuff can float around in space, I have to clean the walls, ceiling, and floor. I have to disassemble the small room and vacuum out the vents—those are especially gross, as they are covered with a year of dust. I also hide a plastic roach for my successor Jeff Williams to find.

Amiko tells me she's had someone come by to check on the pool and hot tub—the pool heater broke partway through my mission and she hadn't noticed until she started putting everything in order for my return. She knows I have been thinking about how great it will be to jump in the pool. She asks me to send a list of the things I want her to have on hand when I get back. Typing the list makes me think about home even more: the sheets of my bed, the shower, the pool and hot tub in the backyard. I've spent this entire year trying not to long for home, and now I'm putting myself there deliberately. It feels very strange.

I email her a list:

Subject: Stuff I Want at Home
Gatorade (the old-school green kind)
Dogfish Head 60 Minute India Pale Ale
And a six-pack of Miller High Life (remember I said I had a craving for that)
Green seedless grapes
Strawberries
Salad stuff
Cabernet
La Crema Chardonnay
Bottled water

Often when I do interviews and press events from space, I'm asked what I miss about Earth. I have a few answers I always reach for that make sense in any context: I mention rain, spending time with my family, relaxing at home. Those are always true. But throughout the day, from moment to moment, I'm aware of missing all sorts of random things that don't even necessarily rise to the surface of my consciousness.

I miss cooking. I miss chopping fresh food, the smell vegetables give up when you first slice into them. I miss the smell of the unwashed skins of fruit, the sight of fresh produce piled high in grocery stores. I miss grocery stores, the shelves of bright colors and the glossy tile floors and the strangers wandering the aisles. I miss people. I miss the experience of meeting new people and getting to know them, learning about a life different from my own, hearing about things people experienced that I haven't. I miss the sound of children playing, which always sounds the same no matter their language. I miss the sound of people talking and laughing in another room. I miss rooms. I miss doors and door frames and the creak of wood floorboards when people walk around in old buildings. I miss sitting on my couch, sitting on a chair, sitting on a bar stool. I miss the feeling of resting after opposing gravity all day. I miss the rustle of papers, the flap of book pages turning. I miss drinking from a glass. I miss setting things down on a table and having them stay there. I miss the sudden chill of wind on my back, the warmth of sun on my face. I miss showers. I miss running water in all its forms: washing my face, washing my hands. I miss sleeping in a bed—the feel of sheets, the heft of a comforter, the welcoming curve of a pillow. I miss the colors of clouds at different times of day and the variety of sunrises and sunsets on Earth.

I also think about what I'll miss about this place when I'm back on Earth. It's a strange feeling, this nostalgia in advance, nostalgia for things I'm still experiencing every day and that often, right now, annoy me. I know I will miss the friendship and camaraderie of the fourteen people I have flown with on this yearlong mission. I'll miss the view of Earth from the Cupola. I know I will miss the sense that I'm surviving by my wits, the sense that life-threatening challenges could come along and that I will rise to meet them, that every single thing I do is important, that every day could be my last.

PACKING UP to leave space is strange. A lot of stuff goes in the trash, which means stowing it in the Cygnus that will burn up in the atmosphere later this month. I throw out a lot of unused clothes—my challenge to myself to use as few clothes as possible has been a success, and there is a duffel bag's worth of T-shirts, sweatshirts, underwear, socks, and pants left over to prove it.

On the weekend, I find the time to take pictures of a bunch of stuff people have asked me to bring—T-shirts, hats with logos, photographs, artwork, jewelry. I gather it all up and take it to the Cupola. As I open the shutters, I catch a glimpse of tawny sand, and I instantly know from the color and texture exactly where we are above the planet: the Somali plains just north of Mogadishu. In one way it's satisfying to feel like I know the planet with such intimacy. In another way, it makes me feel like I've definitely been up here too long.

One by one, I take the items I've brought up here for people and float them against the backdrop of the Earth to snap a picture of each. It's not hard or even that time-consuming, but it's the kind of thing I never felt like doing and that could always be put off until later . . . until now.

There is another thing I wanted to do that I haven't quite found the right time for. I've been thinking about the whole arc of my life that brought me here, and I always think about what it meant to me to read *The Right Stuff* as a young man. I feel certain that I wouldn't have done any of the things I have if I hadn't read that book—if Tom Wolfe hadn't written it. On a quiet Saturday afternoon, I call Tom Wolfe to thank him. He sounds truly amazed to hear from me. I tell him we're passing over the Indian Ocean, how fast we're going, how our communication system works. We talk about books and about New York and about what I plan to do first when I get back (jump into my swimming pool). We agree to have lunch when I'm back on Earth, and that's now one of the things I'm looking forward to most.

On February 29, 2016, I hand over command of the International Space Station to Tim Kopra. Tomorrow I will leave the station and return to Earth.

20

March 1, 2016

Dreamed I was doing a spacewalk with my brother. At first we went outside in our normal clothes, because you could do that if it was a short period of time. Then we went inside and he put on an American spacesuit and I put on a Russian one, the Orlan. I liked the Orlan suit, but I was concerned that I had not trained in it. We went back out of the airlock to find the outside of the space station covered in snow, like a winter wonderland.

THE SIX OF US are gathered in the Russian segment, having another awkward photo op floating in front of the Soyuz hatch. When it's time, Sergey, Misha, and I each hug Tim, Tim, and Yuri and say our good-byes. They snap pictures of us as we float through the hatch. I know from a great deal of experience that it's an odd feeling to say good-bye from that side, knowing that you will be staying behind in space while your friends return to Earth. After having spent so much time together in such close quarters, we've now closed a door between us that won't open again.

Just before Sergey closes the hatch behind us, Misha turns and reaches through to touch the wall of the space station one last time. He gives it a pat, the way you'd pat a horse. I know he's thinking he might not be here again and he's feeling nostalgia for this place that has meant so much to him.

If the process of getting up to space is violent and uncomfortable, the process of coming back down is even more so. Descending in the Soyuz capsule is one of the most dangerous moments of this year, and it will be one of the most physically grueling. Earth's atmosphere is naturally resistant to objects entering from space. Moving at the high speed of orbit, any object will create friction with the air—enough friction that most objects simply burn up from the heat. This is a fact that generally works to our advantage, as it protects the planet from the many meteoroids and orbital debris that would otherwise rain down unexpectedly. And we take advantage of it when we fill visiting vehicles with trash and then set them loose to burn up in the atmosphere. But it's also what makes a return from space so difficult and dangerous. The three of us must survive a fall through the atmosphere that will create temperatures up to three thousand degrees and up to 4 g's of deceleration. The atmosphere seems designed to kill us, but the Soyuz capsule, and the procedures we go through, are designed to keep us alive.

The return to Earth will take about three and a half hours, with many steps we must get through successfully. After pushing away from the station, we will fire the engine to slow us slightly and ease our way into the upper layers of the atmosphere at just the right speed and angle to start our descent. If our approach is too steep, we could fall too fast and be killed by excessive heat or deceleration. If it's too shallow, we could skip off the surface of the atmosphere like a rock thrown at a still lake, only to later enter much more steeply, likely with catastrophic consequences. Assuming our deorbit burn goes as planned, the atmosphere will do most of the work of slowing us down, while the heat shield will (we hope) keep the temperatures from killing us, the parachute will (we hope) slow our descent once we are within ten kilometers, and then the soft landing rockets will (we hope) fire to further slow our descent in the seconds before we hit the ground. Many things need to happen perfectly or we will be dead.

Sergey has already spent days stowing the cargo we will be bringing with us on the Soyuz—our small packages of personal items, water samples, blood and saliva for the human studies. We pack up some trash to be disposed of in the habitation module of the Soyuz, and I include the head of the gorilla suit, since I don't want to be held responsible for any future Space Gorilla antics. Most of the storage space in the capsule is devoted to things we hope we never have to use: the radio, compass, machete, and cold-weather survival gear in case we land off course and must wait for the rescue forces.

Because our cardiovascular systems have not had to oppose gravity all this time, they have become weakened and we will suffer from symptoms of low blood pressure on our return to Earth. One of the things we do to counteract this is fluid loading—ingesting water and salt to try to increase our plasma volume before we return. The Russians and the Americans have different philosophies about the best fluid-loading protocols. NASA gives us a range of options that include chicken broth, a combination of salt tablets and water, and Astro-Ade, a rehydration drink developed specifically for astronauts. The Russians prefer more salt and less liquid, in part because they prefer not to use the diaper during reentry. Having figured out what worked for me on my previous flights, I stick to drinking lots of water and wearing the diaper.

I struggle into my Sokol suit, which is even harder to get into here than it was in Baikonur, where gravity kept things still and I had suit technicians to help me. We used the suits once when we relocated the Soyuz before Gennady left, and I put mine on again a few days ago for the fit check—other than that, it's been waiting for me patiently in the habitation compartment of the Soyuz for a year. As I pull the neck ring up over my head, I try to remember the day I put this suit on for launch, a day when I'd eaten fresh food for breakfast, had taken a shower, and had gotten to see my family. I also saw a lot of other people that day, people everywhere—hundreds altogether, some of them strangers I'd never seen before and would never see again. That is the part that seems strangest now. Everything about that day seems distant to me, like a movie I saw once about someone else.

I'm preparing to climb into the capsule for the ride home, contemplating packing myself into that tiny space again. We float into the center section of the Soyuz, the descent capsule, one by one. First Misha squeezes his tall frame in, closing the hatch partially behind him in order to struggle into the left seat. Misha opens the hatch so I can float down; then I squeeze myself past the hatch, hoping that none of the hardware on my suit scratches up the hatch seal. I get into the center seat, close the hatch to get it out of the way, then awkwardly shimmy myself over to the right seat. Once I'm in, I open the hatch again, and Sergey settles himself into the center seat. We sit with our knees pressed up to our chests.

We are in the seat liners that were custom molded to fit our bodies, and they are more important now than they were on launch day. We will go from 17,500 miles per hour to a hard zero in less than thirty minutes, and the seats, along with many other parts of the Soyuz, must work as designed to keep us on the winning side of a battle against the forces of nature. We strap ourselves in as best we can using the five-point restraints, easier said than done when the straps are floating around us and any tiny force pushes us away from the seats. It's hard to get secured very tightly, but once we are hurtling toward Earth, the full force of deceleration will crush us down into our seats, making it easier to fully tighten our straps.

A command from mission control in Moscow opens the hooks that hold the Soyuz to the ISS, and soon after, spring-force plungers nudge us away from the station. Both of these processes are so gentle that we don't feel or hear them. We are now moving a couple of inches per second relative to the station, though still in orbit with it. Once we are a safe distance away, we use the Soyuz thrusters to push us farther from the ISS.

Now there is more waiting. We don't talk much. This squashed position creates excruciating pain in my knees, as it always has, and it's warm in here. A cooling fan runs to circulate air within our suits, a low comforting whirr, but it's not enough. I remember sitting in the right-hand seat of a different Soyuz, remarking to Misha that our lives without fan noise were over. That seems so long ago. Now, I can't remember what it's like to be in silence, and I yearn to experience it again.

I find it hard to stay awake. I don't know if I'm tired just from today or from the whole year. Sometimes you don't feel how exhausting an experience has been until it's over and you allow yourself to stop ignoring it. I look over at Sergey and Misha, and their eyes are closed. I close mine too. The sun rises; forty-five minutes later, the sun sets.

When we get word from the ground that it's time for the deorbit burn, we are instantly, completely awake. It's important to get this part right. Sergey and Misha execute the burn perfectly, a four-and-a-half-minute firing of the braking engine, which will slow the Soyuz by 300 miles per hour. We are now in a twenty-five-minute free fall before we slam into Earth's atmosphere.

When it's time to separate the crew module—the tiny, cone-shaped capsule we are sitting in—from the rest of the Soyuz, we hold our breaths. The three modules are exploded apart. Pieces of the habitation module and instrumentation compartment fly by the windows, some of them striking the sides of our spacecraft. None of us mentions it, but we all know that it was at this point in a Soyuz descent in 1971 that three cosmonauts lost their lives. A valve between the crew module and the orbital module opened during separation, depressurizing the cabin and asphyxiating the crew. Misha, Sergey, and I wear pressure suits that would protect us in the case of a similar accident, but this moment in the descent sequence is still one we are glad to put behind us.

We feel gravity begin to return, first slowly, then with a vengeance. Soon everything is oddly heavy, too heavy—our checklists, our arms, our heads. My watch feels heavy on my wrist, and breathing gets harder as the g forces clamp down on my trachea. The capsule heats up, and flaming pieces of the heat shield fly by the window as it's scorched black.

We hear the wind noise building as the thick air of the atmosphere rushes past the capsule, a sign that the parachute will soon be deployed. This is the only part of reentry that is completely automated, and we concentrate on the monitor, waiting for the indicator light to show that it worked. It won't be long, maybe only a second or two, before we feel the jerk of the parachute, but we watch anyway. Everything now depends on one parachute, manufactured in an aging facility outside Moscow by similarly aging workers using quality standards inherited from the Soviet space program. After all I've experienced this year—

the long days, the grueling spacewalks, living through the missed birthdays and celebrations, the struggles personal and professional—everything depends on that parachute. We are falling at the speed of sound. We fall and wait and watch.

The chute catches us with a jerk, rolling and buffeting our capsule crazily through the sky. I've heard this experience compared to a train accident followed by a car accident followed by falling off your bike. I've described it myself as the sensation of going over Niagara Falls in a barrel, while on fire. In the wrong frame of mind this would be terrifying, and from what I've heard some people who have experienced it have been terrified. But I love it. It's like a carnival ride on steroids.

Misha's checklist comes loose from its tether and flies at my head. I reach up and grab it out of the air left-handed. The three of us look at one another with amazement.

"Left-handed Super Bowl catch!" I shout, then quickly realize Sergey and Misha might not know what the Super Bowl is. This is not only a moment to revel in my athleticism; it's also a good indication that the motion of the Soyuz must not be as crazy as it seems to us—a lot of the perceived motion is our vestibular systems overreacting to the force of gravity.

After all the tumult of the descent, the minutes we spend drifting at the whim of the parachute are oddly calm. Later I will see a photograph taken of our Soyuz dangling under the white-and-orange parachute against the backdrop of a fluffy blanket of clouds. The heat shield is jettisoned, pulling off the burned window coverings. Sunlight streams in the window at my elbow as we watch the ground come closer and closer.

From their position in helicopters nearby, the rescue forces count down over the comm system the distance to go until landing.

"Open your mouth," a voice reminds us in Russian. If we don't keep our tongues away from our teeth, we could bite them off on impact. When we are only five meters from the ground, the rockets fire for the "soft" landing (this is what it's called, but I know from experience that the landing is anything but soft). I feel the hard crack of hitting the

Earth in my spine. My head bounces and slams into the seat, the sensation of a car accident. We are down. We have landed with the hatch pointing straight up rather than on one side, which is rare. We will wait a few minutes longer than usual while the rescue crew brings a ladder to extract us from the burned capsule.

When the hatch opens, the Soyuz fills with the rich smell of air and the bracing cold of winter. It smells fantastic. We bump fists.

After Sergey and Misha get out of the capsule, I'm surprised to find that I can unstrap myself, pull myself out of my seat, and reach the hatch overhead, despite the fact that gravity feels like a crushing force. I remember coming back from STS-103 after only eight days and feeling like I weighed a thousand pounds. Now, with a little help from the rescue forces, I pull myself entirely out of the capsule to sit on the edge of the hatch and take in the landscape all around. The sight of so many people—maybe a couple hundred—is startling. It feels indescribably strange to see more than a handful of people at a time, and the sight is overwhelming. I pump my fist in the air. I breathe, and the air is rich with a fantastic sweet smell, a combination of charred Soyuz and honeysuckle. The Russian space agency insists on having the rescue crew help us down from the capsule and deposit us into nearby camp chairs for examination by doctors and nurses. We follow the Russians' rules when we travel with them, but I wish they would let me walk away from the landing. I feel sure I could. My flight surgeon Steve Gilmore is there, and I'm reminded of what his medical care and friendship have meant to me—over the years, he and other flight surgeons have worked tirelessly to keep me on flight status and kept me flying safely when it would have been easier to declare me unqualified. I notice Chris Cassidy, the chief astronaut, and my friend Joel Montalbano, the deputy ISS program manager. Near Sergey and Misha, I recognize Sergey's father, a former cosmonaut, and Valery Korzun. In the distance, I see the rescue force troops, some of whom I first met in Russa in 2000 during winter survival training and whose dedication I have come to appreciate and rely on. I notice Misha smiling and waving at them, and I'm certain he's thinking of his father, who was once one of them.

Chris hands me a satellite phone. I dial Amiko's cell—I know she'll be at mission control in Houston along with Samantha (Charlotte watched from home in Virginia Beach), my brother, and close friends watching the live feed on the huge screens.

"How was it?" Amiko asks.

"It was fucking medieval," I say. "But effective."

I tell her I feel fine. If I were on the first crew to reach the surface of Mars, just now touching down on the red planet after a yearlong journey and a wild-hot descent through its atmosphere, I feel like I would be able to do what needed to be done. One of the most important questions of my mission has been a simple yes or no: could you get to work on Mars? I wouldn't want to have to build a habitat or hike ten miles, but I know I could take care of myself and others in an emergency, and that feels like a triumph.

I tell Amiko I'll see her soon, and for the first time in a year that's true.

Epilogue: Life on Earth

I've been asked often what we learned from my year in space. I think sometimes people want to hear about one profound scientific discovery or insight, something that struck me (or the scientists on the ground) like a cosmic ray through my brain at some climactic moment during my mission. I don't have anything like that to offer. The mission that I prepared for was, for the most part, the mission I flew. The data is still being analyzed as I write this, and the scientists are excited about what they are seeing so far. The genetic differences between my brother and me from this year could unlock new knowledge, not only about what spaceflight does to our bodies but also about how we age here on Earth. The Fluid Shifts study Misha and I did is promising in terms of improving astronauts' health on long missions. The studies I did on my eyes—which don't seem to have degraded further during this mission—could help solve the mystery of what causes damage to astronauts' vision, as well as helping us to understand more about the anatomy and disease processes of the eye in general.

Results and scientific papers will continue to emerge over years and decades based on the four hundred experiments we conducted over the year. Misha and I were a sample size of only two—we need to see many more astronauts stay in space for longer periods of time before we can draw conclusions about what we experienced. I do feel as though I've made discoveries—it's just that those discoveries can't entirely be separated from what I've learned from my other missions in space, other periods of my life, other challenges, other lessons.

As much as I worked on scientific experiments, I think I learned at least as much about practical issues of how to conduct a long-range exploration mission. This is what crew members on ISS are always doing—we are not just solving problems and trying to make things better for our own spaceflights, but also studying how to make things better for the future. So even the smallest decisions I made or negotiations I undertook with the ground were directed toward larger questions of resource management. And the larger struggles of my mission—most notably, CO_2 management and upkeep of the Seedra—will have a larger impact on future missions on the space station and future space vehicles. NASA has agreed to manage CO_2 at a much lower target level, and better versions of carbon dioxide scrubbers are being developed that will one day replace the Seedra and make life better for future space travelers, and I'm thankful for that.

Personally, I've learned that nothing feels as amazing as water. The night my plane landed in Houston and I finally got to go home, I did exactly what I'd been saying all along I would do: I walked in the front door, walked out the back door, and jumped into my swimming pool, still in my flight suit. The sensation of being immersed in water for the first time in a year is impossible to describe. I'll never take water for granted again. Misha says he feels the same way.

I've been assigned to a spaceflight or in training for one practically nonstop since 1999. It will be an adjustment to no longer be planning my life this way. I have a chance to reflect on what I've learned.

I've learned that I can be really calm in bad situations. I've known this about myself since I was a kid, but it has definitely been reinforced.

I've learned to better compartmentalize, which doesn't mean forgetting about feelings but instead means focusing on the things I can control and ignoring what I can't.

I've learned from watching my mother train to become a police officer that small steps add up to giant leaps.

I've learned how important it is to sit and eat with other people. While I was in space, I saw on TV one day a scene of people sitting down to eat a meal together. The sight moved me with an unexpected yearning. I suddenly longed to sit at a table with my family, just like the people on the screen, gravity holding a freshly cooked meal on the table's surface so we could enjoy it, gravity holding us in our seats so we could rest. I had asked Amiko to buy a dining room table; she did, and sent me a picture of it. Two days after landing, I was sitting at the head of the new table, a beautiful meal my friend Tilman had sent over spread out on it, my family gathered around me. Amiko, Samantha, Charlotte, Mark, Gabby, Corbin, my father. I could see them all without moving my head. It was just how I'd pictured it. At one point in the after-dinner conversation, Gabby pointed urgently at Mark, then me, back and forth, back and forth. She was pointing out that Mark and I were both making exactly the same gesture, our hands folded on top of our heads. I've learned what it means to be together with family again.

I've learned that most problems aren't rocket science, but when they are rocket science, you should ask a rocket scientist. In other words, I don't know everything, so I've learned to seek advice and counsel and to listen to experts. I've learned that an achievement that seems to have been accomplished by one person probably has hundreds, maybe even thousands, of people's minds and work behind it, and I've learned that it's a privilege to be the embodiment of that work.

I've learned that Russian has a more complex vocabulary for cursing than English does, and also a more complex vocabulary for friendship.

I've learned that a year in space contains a lot of contradictions. A year away from someone you love both strains the relationship and strengthens it in new ways. I've learned that climbing into a rocket that may kill me is both a confrontation of mortality and an adventure that makes me feel more alive than anything else I've ever experienced. I've learned that this moment in American spaceflight is a crossroads where we can either renew our commitment to push farther out, to build on our successes, to keep doing harder and harder things—or else lower our sights and compromise our goals.

I've learned that grass smells great and wind feels amazing and rain is a miracle. I will try to remember how magical these things are for the rest of my life.

I've learned that my daughters are remarkable and incredibly resilient people, and that I have missed a piece of each of their lives that I can never get back.

I've learned that following the news from space can make Earth seem like a swirl of chaos and conflict, and that seeing the environmental degradation caused by humans is heartbreaking. I've also learned that our planet is the most beautiful thing I've ever seen and that we're lucky to have it.

I've learned that voluntary spinal taps are not much fun.

I've learned a new empathy for other people, including people I don't know and people I disagree with. I've started letting people know I appreciate them, which can sometimes freak them out at first. It's a bit out of character. But it's something I'm glad to have gained and hope to keep.

I TOLD my flight surgeon Steve I felt well enough to get right to work immediately upon returning from space, and I did, but within a few days I felt much worse. This is what it means to have allowed my body to be used for science. I will continue to be a test subject for the rest of my life.

A few months later, I felt distinctly better. I will continue to participate in the Twins Study as Mark and I age. Science is a slow-moving process, and it may be years before any great understanding or breakthrough is reached from the data. Sometimes the questions science asks are answered by other questions. This doesn't particularly bother me—I will leave the science up to the scientists. For me, it's worth it to have contributed to advancing human knowledge, even if it's only a step on a much longer journey.

I've been traveling the country and the world talking about my experiences in space. It's gratifying to see how curious people are about my mission, how much children instinctively feel the excitement and wonder of spaceflight, and how many people think, as I do, that Mars is the next step.

In the summer, my father was diagnosed with throat cancer and began receiving radiation therapy. In October, he became much more ill. One evening, Amiko got a phone call from him, which wasn't unusual. He had depended on her support a great deal while I was in space, and they had continued to talk often. But that day, he didn't want anything in particular.

"I just wanted to let you know how much I love you, sweetheart," he told her. "I'm just so glad you and Scott have each other. You've accomplished so much together, and all the stuff you've been through—it was all worth it." Amiko thought this was out of character for him, but she said he sounded much better than he had in a while. A few days later, he took a turn for the worse, and while Mark, Amiko, and I were all out of the country, he died in the intensive care unit with my daughter Samantha by his side, four and a half years after my mother. I was grateful Samantha could be there with him.

I'm convinced he lived to see my mission through and to celebrate my return. It was a big deal to him to support Mark and me and to celebrate our accomplishments, and he was proud of all of his granddaughters, whom he adored. Like most people, he had mellowed with age, and we had a much improved relationship toward the end of his life.

In my computer, I have a file of all the images my crewmates and I took on the International Space Station during the time I was there. When I'm trying to remember some detail from the mission, sometimes I click through them. It can be overwhelming, because there are so many of them—half a million—but often a picture of a specific person on a specific day will bring back a flood of sense memory, and I will suddenly remember the smell of the space station, or the laughter of my crewmates, or the texture of the quilted walls inside my CQ.

One night, I click through the images late at night after Amiko is asleep: an image of Misha and Sergey in the Russian service module, smiling, getting ready for a Friday night dinner; an image of Samantha Cristoforetti running on the treadmill on the wall, grinning; an image of a purple-and-green aurora that I took in the middle of the night;

an image of the eye of a hurricane taken from above; an image of a dirty filter vent just before I threw it away, containing a snarl of dust, lint, and one really long blond hair that must have come from Karen Nyberg, who left the station more than a year before I got there; a series of images of the connectors on the Seedra Terry and I took while we were in the process of fixing it so the ground could see how it was looking; an image of an iPad floating in the Cupola displaying a snapshot of a newborn baby I don't know, majestic cloud formations visible below; an image of Tim Peake preparing his spacesuit for his first spacewalk, the Union Jack visible on the shoulder of the suit, a huge boyish grin on his face; a picture of Kjell flying through the U.S. lab like Superman; a picture of Gennady and me chatting in Node 1, just enjoying the moment and each other's company. A year is made up of a million of these, and I could never have captured them all.

One image that doesn't exist in my computer but that I will always remember is the view from the Soyuz window as Sergey, Misha, and I backed away from the International Space Station. As well as I know the inside of the station, I've only seen the outside a handful of times. It's a strange sight, glinting in the reflected sunlight, as long as a football field, its solar arrays spread out more than half an acre. It's a completely unique structure, assembled by spacewalkers flying around the Earth at 17,500 miles per hour in a vacuum, in extremes of temperature of plus and minus 270 degrees, the work of fifteen different nations over eighteen years, thousands of people speaking different languages and using different engineering methods and standards. In some cases the station's modules never touched one another while on Earth, but they all fit together perfectly in space.

As we backed away, I knew I would never see it again, this place where I'd spent more than five hundred days of my life. We will never have a space station like this again in my lifetime, and I will always be grateful for the part I've played in its life. In a world of compromise and uncertainty, this space station is a triumph of engineering and cooperation. Putting it into orbit—making it work and keeping it working—is the hardest thing that human beings have ever done, and it stands as proof that when we set our minds to something hard, when we work together, we can do anything, including solving our problems here on Earth.

I also know that if we want to go to Mars, it will be very, very difficult, it will cost a great deal of money, and it may cost human lives. But I know now that if we decide to do it, we can.

Acknowledgments

Amiko said to me once, "Teamwork makes the dream work," and spaceflight is the biggest team sport there is, so spending any amount of time in space takes the support and collaboration of thousands of people. From the instructors who train us to the flight controllers and flight directors working in mission control to my friends and family keeping me connected to my life on Earth—there isn't enough space in this book to thank them all, so one collective "thank you" will have to suffice.

Above all, I have to recognize my partner—and now fiancée—Amiko Kauderer. I hope the pages of this book make clear what it meant to me that she was with me day by day throughout this journey, experiencing together its challenges and triumphs and its highs and lows. I've tried to express what a crucial role she's played in this mission's success, but words can never express the role she has played in my life these last eight years. Thank you, Amiko.

My kids, Samantha and Charlotte, have sacrificed much for their dad. From missed birthdays and holidays to the general disruption of their lives, accepting the inherent risks of spaceflight and sharing their dad with the world. They were brave, adaptable, and resilient. I appreciate and am proud of how you handled it all with strength and grace. Thank you.

My brother, Mark, has been by my side since my birth, challenged and supported me throughout our lives. Having also flown in space, he understood the thrill, the trials, and the hardships of this journey. His support and counsel I've come to rely on and much appreciate. Thank you.

My parents endured the emotional toll of watching their sons launch into space and await our safe return to Earth—a total of seven times for my mother, Patricia, and eight times for my father, Richard. Thank you to my mother also for showing me by her example what it took to achieve a lofty goal.

My ex-wife, Leslie, lent her willing support, accepting the role of a full-time single parent, ensuring our daughters were safe and cared for back on Earth each time I went to work off the planet. Thank you.

Writing a book is a team effort as well. This is my first experience writing a book, but it was also the first time my collaborator Margaret worked on someone else's book. Despite this, the experience couldn't have been an easier collaborative effort. From the beginning, Margaret showed herself as trustworthy in not only maintaining confidentiality but also by allowing me to open myself and explore my own emotions, which helped bring these personal stories to life. Thanks, Margaret, for helping me through this process and thanks for your friendship.

Our editor, Jonathan Segal, was also critical to the process and final product. Thank you, Jonathan. I also need to thank my literary agent, Elyse Cheney, for not only inking this deal with the publisher but also for being a mentor and friend.

My flight surgeon, Dr. Steve Gilmore, deserves a special recognition for taking care of my health in space and on Earth for many years and for providing critical insight to much of the medically related content.

I have to thank some of the people who offered their perspectives on the experiences I describe and allowed me to tell their stories. Many people helped by filling in details, offering input on drafts of the book, and helping out in other ways big and small. Thank you to Bill Babis, Chris Bergin, Dr. Steve Blackwell, Beth Christman, Paul Conigliaro, Samantha Cristoforetti, Dr. Tracy Caldwell Dyson, Tilman Fertitta, Steve Frick, Dr. Bob Gibson, Marco Grob, Ana Guzman, Martha Handler, Dr. Elena Hansen, Brooke Heathman, Christopher Hebert, Giselle Hewitt, Dr. Al Holland, Akihiko Hoshide, Bill Ingalls, Omar Izquierdo, Dr. Smith Johnston, Dr. Jeff Jones, Bob Kelman, Sergey Klinkov,

Nathan Koga, Mike Lammers, Dr. Kjell Lindgren, Dr. Gioia Massa, Dr. Megan McArthur, Dr. Brian Miles, Rob Navias, Dr. James Picano, Dr. Julie Robinson, Jerry Ross, Tom Santangelo, Daria Shcherbakova, Kirk Shireman, Scott Stover, Jerry Tarnoff, Robert Tijerina, Terry Virts, Sergey Volkov, Dr. Shannon Walker, Dr. Liz Warren, Doug Wheelock, and Dr. Dave Williams.

And finally, I have to thank Tom Wolfe for his early inspiration. I truly believe if I had not read *The Right Stuff* as an eighteen-year-old, I would not have written this book or had the privilege of flying in space.

Index

Page numbers in *italics* refer to illustrations.

ILLUSTRATION CREDITS

Pages 2–3: Nathan Koga; pages 47, 78, 88, 225, and 292: NASA; pages 70 and 74: NASA/Scott Kelly

Insert 1

Page 1: NASA (top and bottom); page 2: Scott Kelly (top and middle), U.S. Navy (bottom); page 3: NASA; page 4: NASA (top and bottom); page 5: NASA (top and bottom); page 6: NASA (top and bottom); page 7: NASA (top and bottom); page 8: Scott Kelly (top left), Stephanie Stoll/NASA (top right), NASA (bottom)

Insert 2

Page 1: NASA/Bill Ingalls (top), NASA (bottom); page 2: NASA (top), NASA/Bill Ingalls (bottom); page 3: NASA/Bill Ingalls (top), NASA (bottom); page 4: NASA (top), NASA/Bill Ingalls (bottom); page 5: NASA/Bill Ingalls (top and bottom); page 6: NASA/Bill Ingalls (top), NASA (bottom); page 7: NASA (top and bottom); page 8: Scott Kelly/NASA (top), NASA (bottom); page 9: NASA (top), NASA/Scott Kelly (bottom); page 10: NASA (top), NASA/Scott Kelly (bottom); page 11: NASA/Scott Kelly (top), NASA (bottom); page 12: NASA/Scott Kelly (top and bottom); page 13: NASA/Scott Kelly (top and bottom); page 14: NASA/Bill Ingalls (top and bottom); page 15: NASA/Bill Ingalls; page 16: NASA

About the Author

Scott Kelly is a retired astronaut and International Space Station commander. He recently returned from his #YearInSpace to break all records and become the longest-serving NASA astronaut in space.

Scott Kelly's contribution to his field is immeasurable and his experience utterly unique. *Endurance* brings it to life for those of us who may never make it beyond the Earth's atmosphere.